ECOLOGY OF PLANT-PARASITIC NEMATODES

ECOLOGY OF PLANT-PARASITIC NEMATODES

DON C. NORTON

Department of Botany and Plant Pathology
Iowa State University, Ames

A WILEY-INTERSCIENCE PUBLICATION

JOHN WILEY & SONS, New York • Chichester • Brisbane • Toronto

Library of Congress Cataloging in Publication Data:

Norton, Don C.
    Ecology of plant-parasitic nematodes.

    "A Wiley-Interscience publication."
    Bibliography: p.
    Includes index.
    1. Plant nematodes—Ecology.   2. Nematoda—
Ecology.   I. Title.

SB998.N4N64   632'.6'5182   78-1052
ISBN 0-471-03188-7

Printed in the United States of America

10 9 8 7 6 5 4 3 2 1

To the memory of

JOHN N. WOLFE

Scientific results of importance in all branches of research can be attained only by persistent observations during a lengthened sojourn in these regions...

FRIDTJOF NANSEN
Farthest North

PREFACE

This book is about the ecology of nematodes that parasitize plants. My approach is via the habitat, and my theme is "Propagules are everywhere; the environment selects," certainly not original with me.

Nematode ecology has not had the traditions that other disciplines have enjoyed. Economic necessities and concomitant funding have directed the bulk of research on nematodes toward dealing with food and fiber crops. Although this book is not concerned directly with dollars-and-cents agriculture, I hope that it will make a contribution toward that end. Consequently, this treatment is not about disease per se but is an attempt to examine one small facet of a group of organisms that under some conditions cause disease, damage, or whatever one wishes to call it. Although parasitism has much to do with the contents, pathogenicity for the most part will be omitted. Wallace (1973) has written from a different approach that reflects the ecology of nematodes relative to disease.

Plant nematodes are relatively unknown except by professional nematologists and in some areas of agriculture. Professional plant ecologists may hardly know that nematodes exist, yet nematodes are part of the ecosystem and should be treated as such. Although we probably would not want to change the nematode's role in the natural ecosystem if we could, we cannot help but wonder what the world's vegetational patterns would be if all parasitic nematodes were eliminated. They are a part of the natural world, and we have grown up with them. But certainly a knowledge of their presence and their role in natural ecosystems will provide a better understanding of the dynamics of these systems. Relatively little has been accomplished but there are indications of growing interest in these aspects in the present environmental age. I have tried to combine aspects of agricultural and natural ecosystems in my book, although there is far more information on the former than the latter.

The presence of and damage caused by nematodes are insidious. In contrast with some plant pests, nematodes are not seasonal in their appearance. Nematodes parasitize roots from the time that the roots emerge from the seed until the roots die. Some nematode species parasitize other plant parts also. The nematode populations attained will depend on many factors, but the environment, including the host, is the most important. In this book nematodes are treated from the broad ecological viewpoint emphasizing the habitat and the effects of the microenvironment. The material is also presented with the realization that much has been done in some categories but little in others. Only a simplified introduction to population ecology is provided here. We are just beginning the "modern" ecology in nematology and the techniques have mostly been borrowed from other disciplines to which the reader is referred. Plant nematology as a whole is far behind other disciplines in ecological matters.

Many ecological data are scattered in the literature, most of them empirical. Some data are assembled here for a general perspective, but also there is enough documentation to make the reader aware of the diversities and to provide sufficient background for delving deeper into the subject. There are shortcomings imposed upon the treatment by necessity. For example, research on the saprobes has not progressed sufficiently as they relate to plant nematodes and must await later treatment. Another omission, but one for which there is a large volume of data, is alteration of nematode populations by chemical means. The data, most of them empirical, are so voluminous that a sizable volume on the subject only awaits the effort to write it.

Most generalizations at the beginning of each chapter were obtained from John Wolfe during his course in plant ecology. I make no attempt to defend them relative to nematode ecology but I thought that they were pertinent.

One final word on distribution. My emphasis on this, as well as other aspects, has been placed in the United States, a pragmatic if not an ideal approach. Known distribution patterns and reasons behind them are still in their infancy. To interpret biogeography, and to have adequate knowledge in the agricultural and nonagricultural ecosystems and the energy flows in these systems, we need ecological surveys and experimentation. Every mountain range, every plain, marsh, desert, wood, and prairie, and every agricultural type of habitat should be studied as intensively for nematodes as they have been for other plants and animals. In this respect, the descriptive phase has hardly begun, to say nothing of the functional phase. Nematology is a young science; we are probably at the lower end of the logistic curve. There are ever so many discoveries awaiting the investigator. Facts are needed not just for the sake of having facts but as building blocks of knowledge of the ecosystem, something that extends beyond the pragmatism of agriculture, as important as that is. The presence of nematodes in the ecosystem, agricultural or other, be it *Heterodera rostochiensis* in a potato field in Germany or a relatively obscure *Tylenchus* in a woodland, is involved in the interrelationships of all living things. We have learned many bits and pieces concerning nematodes, but we have not really understood their role in the ecosystem. We have a long way to go.

Many people have assisted me in various ways, and I am very grateful to them. Special thanks go to Victor Dropkin, who read the entire manuscript in an intermediate stage. His encouragement and helpful suggestions are greatly appreciated. L. R. Frederick, J. W. Potter, and J. L. Townshend read selected chapters, and their corrections and comments were most helpful. Special thanks are extended to my wife, Joanne, for helping with clarity and grammatical corrections on nearly every page. Appreciation is expressed to E. J. Brill Publishing Company, Macmillan Journals Ltd., the *New Zealand Journal of Science,* and The Williams and Wilkins Company for permission to use illustrative material. The journal sources are acknowledged in the text. To the many persons who furnished illustrative material or otherwise spent time on my behalf, I will be eternally grateful. In spite of such assistance, the errors and opinions are my own.

DON C. NORTON

Ames, Iowa
February 1978

CONTENTS

ECOLOGY OF PLANT-PARASITIC NEMATODES

ASPECTS OF GEOGRAPHICAL DISTRIBUTION

Every biotic community of the present is a continuation of communities of the past.

Communities that developed under one set of climatic conditions may persist in special habitats long after these general conditions have passed away.

As for all forms of life, there has to have been a beginning for nematodes, and these beginnings are obscure. The known fossil record is extremely limited. Phylogenetic relationships must be surmised from contemporary species. Obviously, nematodes had to evolve from other forms of life, either by ascendancy or retrogression, before a fossil record could appear. Evolutionary theories relative to nematodes have been reviewed by Chitwood and Chitwood (1950), Maggenti (1971), and Paramonov (1962). Although there seems to be general agreement that nematodes parasitic on higher plants were derived from algal feeders, mycophgous nematodes, or predators, there is less agreement as to whether or not the primitive species were terrestrial, including freshwater aquatics, or marine.

THE FOSSIL RECORD ————————————————————————

The ecology and distribtuion of nematodes begins with their phylogenetic origin. Some paleontologists speculate that nematodes originated in the Cambrian period or even in the Precambrian era although a fossil record is unknown. Such speculations are probably correct, because when fossils or their relatives did appear they were sufficiently complicated and diverse that their evolvement must have been underway for some time. Nematodes have few hard structures, and they are less apt to be preserved than some other animals. Nevertheless, fossils of soft organisms are known from the Precambrian era; intensive searching may reveal more of the early nematode fossil record, if such exists.

There is little doubt that free-living forms greatly pre-date the earliest known nematode fossil record. Wavelike tracks found in Eocene strata in Utah (Figure 1.1) are some of the earliest fossils attributed to nematodes because of their striking resemblance to sinusoidal tracks that nematodes make on agar (Moussa, 1969). There are no other animals, fossil or contemporary, known to be capable of making such tracks. Taylor (1935) reviewed the available literature on fossil nematodes up to 1935 and renamed four apparently free-living species found in Oligocene amber near the Baltic Sea.

Theoretically, the first plant-parasitic nematodes could have arisen shortly after plants evolved. Nonseptate fungi are known from the Precambrian era. Pteridophytes were common in the Devonian period, conifers in the Permian, and Angiosperms in the Jurassic period of the Mesozoic era (Darrah, 1960). Land plants and animals are not known in fossils of the

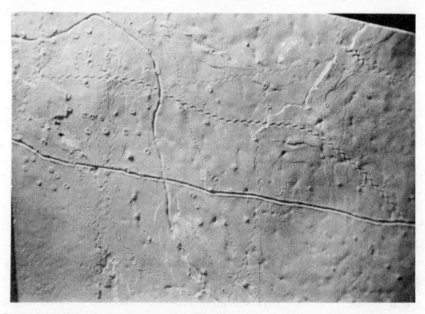

Figure 1.1. Wavelike tracks attributed to nematodes in Eocene rock in Utah. Natural size. (From Moussa, 1969, p. 378.)

Cambrian period (Palmer, 1974); if, as this indicates, life on land was exceedingly sparse, any nematodes existing then were most likely marine. If this is true, tylenchids are probably post-Cambrian even though many nematodes that are generally considered primitive Tylenchida are known fungal feeders. Judging by the parasitism of contemporary nematodes, some phylogentically advanced nematodes such a *Xiphinema* and members of the Criconematinae, known parasites of Gymnosperms, could have existed in the Pennsylvanian and Permian periods when Gymnosperms were abundant. This implies that there were even more ancient forerunners of these advanced nematodes. It can only be conjectured whether or not parasitism of fossil plants by contemporary or fossil nematodes had the same host–parasite relationships that exist today.

Whatever or whenever the origin of nematodes, it is probable that there were many aborted attempts at substrate colonization before colonies were firmly established. Mutations and gene recombinations occur in all directions, and geneticists tell us that relatively few of the mutant offspring are successful. But geologic time is a long time. As habitats changed as a result of natural forces, nematodes species inexorably formed and became established, and in time many became extinct. The progression of ecological events in the past has culminated in the ecological events of today. The more familiar we become with the present relationships of nematodes and plants, the more reliable will be our educated guesses concerning events of the past.

OTHER PREHISTORIC EVENTS

Nematodes probably have had the same patterns and problems of distribution as other organisms. Most soil nematodes are microscopic and lightweight and are probably disseminated readily. They are motile, although this attibute is much less important than passive mobility for any appreciable spread. Once plant-parasitic nematodes evolved, they were subjected to either uniform or highly diverse habitats for varying periods. These habitats, including a diversity of hosts, greatly influenced the rapidity and direction of nematode evolution. Since these events are almost entirely prehistorical, their importance relative to present-day nematode ecology can be deduced only subjectively.

Continental Drift

When we consider nematode distribution and geologic time, we must also consider continental drift. The theory of continental drift is not new, having been proposed by several scientists in recent centuries. In only relatively recent times has it been generally accepted as fact, owing to the convincing evidence of plate tectonics. Acceptance of this theory may resolve some problems relating to nematode distribution, but only tentatively, because we know so little about the geologic record of nematodes or their current distribution. However, assuming that great expanses of water are efficient barriers to terrestrial nematode dispersal and that such barriers are important in species isolation, continental drift may explain, if parallel evolution does not, common occurrences of nematodes in specialized habitats that are now separated by oceans. It implies that these species or their progenitors existed as far back as the early Mesozoic era.

It was only 50 million years ago that India merged with the mainland of Asia, after separating from eastern Africa and Antarctica and traveling across the Tethys Ocean. Were some nematodes, the progeny of which may now be found in Asia where they previously had not occurred, transported by this means? Did some nematodes become extinct due to the adverse environmental shifts encountered? How long do nematode species exist before they evolve to extinction? Are some taxa rather stable, existing today much as they did millions of years ago, as suggested by Ferris et al. (1976)? These questions and problems along with others pose academic exercises that may be basic to our understanding of contemporary nematode distribution. But the record is scanty at best, and many more data are needed before conclusions can be more than highly speculative.

The Pleistocene

Much work has been done concerning the Pleistocene on Quaternary biota (Wright and Frey, 1965). But because of the lack of information about nematodes, inferences must be drawn from studies with other organisms, which is not an ideal situation. As with other organisms, the effect of the Ice Age on nematodes probably was profound during and after the Pleistocene. There is no conclusive evidence that life covered by the ice was able to survive the thousands of years of deep freeze. There is less agreement on the extent and importance of nonglaciated areas as refugia and distribution centers after the ice receded. The nonglaciated area of Wisconsin probably was the most important one in North America, but even this possibility is being reexamined because there is evidence that this Driftless Area was glaciated at an early time. Other possible refugia that have received attention are in Alaska, nunataks in northern New England, and areas along the North Atlantic coast. Life at the glacial border was not static. Also, the ice caused climatic changes far to the south. The eustatic and isostatic changes during the glacial and interglacial periods doubtless effected climatic changes, with consequent changes in the biota. Wolfe (1951) speculates that microclimates south of the glacial border served as plant refugia during these climatic changes, and he argues that they could be important in explaining disjuncts and relicts. If plants could survive in these micro-climates, probably many of their parasites could also.

The impression should not be left, however, that the events associated with the Pleistocene were the major causes of speciation by providing refugia where survival and evolution occurred in isolated habitats. While some relicts or disjuncts of some forms of life might best be explained as survivors from the ice in refugia, much speciation in isolated communities is occurring today, and it has little to do directly with the mantle of ice that left the United States 8000 to 10,000 years ago.

Land bridges connecting North America with Asia or Europe have been cited as possible migration routes for organisms (Frey, 1965). How important these land bridges were in nematode dissemination is a matter of conjecture. The most recent land bridge connecting North America and Asia, now submerged beneath the Bering Strait, was well formed about 18,000 years ago. Evidence indicates that the bridge was a treeless tundra inhospitable to most plant life. While it is possible that some plant-parasitic nematodes migrated over this route, the absence of a more diverse flora undoubtedly prevented many nematodes from becoming established. The possibility of a recent land bridge connecting North America and Europe is more controversial, although there is some biological evidence for its existence. The Davis Strait between Greenland and the mainland probably was a barrier to migration during the Pleistocene, however. Nematologi-cally, this background can be interpreted reasonably well only when more data are obtained on nematode distribution and the timing of species evolu-tion. This knowledge would then have to be weighed with the botanical and geological data upon which the nematologist must largely depend. Obviously, much must be done on nematode geography.

THE RECENT PAST

Recent discoveries (Davidson and Curtis, 1973; Webley, 1974) that cysts of *Heterodera avenae, H. cruciferae, H. galeopsidis* (?), *H. rosii,* and *H. schachtii* were intact after surviving approximately 1000 to 2000 years under minimal oxidation conditions may provide insight to some nematode occurrences in the more recent past. This aspect should be developed further where conditions for survival are favorable.

The Impact of Agriculture

According to historians, agriculture started about 8000 to 12,000 years ago. For a while it was largely nomadic, with natives collecting food in the wild. But as man began to settle and it became increasingly difficult to bring food from afar for an expanding population, he was forced to turn to the more laborious practice of cultivation.

Few historical events, recent or geological, have had such sudden impact on changes in nematode populations as when man first broke the sod or cleared the land for cultivation. Even though the soil is still here, it has become highly modified. A once relatively stable habitat, with probably a relatively stable but shifting and fluctuating nematode fauna, soon became an unstable habitat subject to much wider extremes and greater shifts in properties than hiterhto existed. The catastrophic upheavals of the plow caused more drying of the upper layers of the soil, thus reducing the available moisture, removed a source of food that the nematodes previously had used, and drastically changed the aeration and gas-exchange patterns. With continued cultivation, other changes occurred. New food sources were introduced; the structure of the soil changed; shifts in pH, organic matter, cation exchange, and other soil properties occurred. Brief to long periods of fallow or a nonhost crop created conditions that in many instances exceeded the ecological amplitude of nematodes. It is no wonder that the nematode fauna of cultivated and noncultivated soils are often markedly different, both qualitatively and quantitatively.

The plow is still breaking virgin areas in some localities of the United States, especially in range and forested areas. Most of this has occurred in the United States during the past 100 years, but in Europe and other older inhabited areas the effects of cultivation have been known for centuries. Unfortunately for us, the pioneers were not nematologists, and we can only surmise what probably occurred by studying the presently undisturbed areas, which are increasingly difficult to find. We should take advantage, however, of our position in North America. In addition to the large portion of the continent that was glaciated, a feature common with some other continents, we have the unique situation that it has been relatively undisturbed by cultivation until relatively recent times. We thus have opportunities to study dissemination and distribution in contexts not afforded by some countries that have been in cultivation longer.

Patterns of Distribution

Any discussion of geographical distribution of plant-parasitic nematodes must be met with some trepidation. Sooner or later, distribution maps are constructed, and the resulting pattern, or lack of it, forms an impression in the mind. But, as Wallace (1963) pointed out, world distribution maps are of limited value, because not all areas have been surveyed, and thus to some extent the maps reflect the distribution of nematologists. In addition, areas are surveyed to difficult intensities, depending upon the interests and thoroughness of the investigators. Doubtless, world patterns exist; only future research will fill in the gaps or make them more conspicuous.

Distribtuion maps of species that are controlled by climatic forces are only approximate; sometimes more is read into them than they actually convey. They are approximate because the actual boundaries are seldom known, due to inadequacy of sampling, a formidable task compared with above-ground surveys for macroscopic plants and animals. In addition, and as with other organisms, the boundaries are probably fluctuating according to seasonal weather changes. Favorable climates for a decade or so may allow a species to make inroads into an adjacent area not previously occupied by that nematode, only to be eliminated and perhaps pushed back beyond its previous border by one adverse climatic season. Influences may not be direct, such as those imposed by temperature and moisture. Theoretically, indirect effects of climatic changes may be responsible, such as effects on survival characteristics, physiological changes in the host, or alterations in aggressiveness of competitive species, including parasites and predators. In addition, although distribution maps may indicate general boundaries of a species, they tell nothing of populations. Marginal locations frequently indicate marginal populations, and the most favorable area for nematode activity may be quite distant. As has often been pointed out, distribution and abundance are two aspects of the same idea (Andrewartha, 1970). Also, distribution maps do not reveal the specialized habitats that restrict a nematode to local areas even though the species is distributed widely geographically.

Are the recorded distribution patterns obtained purely a matter of chance, or are they controlled by major geographical and climatic forces? With ample opportunity for dissemination of indigenous species in continuous land masses, and where there are no major geographical barriers such as deserts and mountain ranges, nematodes eventually have come to colonize diverse habitats. Small modifications in habitats may alter the nematode–habitat relationship and may result in speciation and morphological variants of the nematodes present. Relatively slow ecological changes occur in natural areas and probably result in slow changes in nematode community structure. With the agricultural revolution, a new and widespread food source suddenly appeared over vast areas, providing a greater chance for some nematodes to become established and resulting in a more rapid overall rate of dissemination than in noncultivated areas. Brock (1966) contends that historical and geographical factors are of minor importance in microbial ecology and that the environment selects. From some standpoints, the geographical and historical factors such as glaciation have much to do with topography and climatic forces, which in turn may limit nematode establishment.

After examining host ranges and environmental factors, one is almost forced to conclude that geographical isolation plays an important part in the diversification of nematode species. Modern transportation may overcome these barriers, but geographical factors must be considered formidable. Geographical isolation still permeates the reasoning behind the continued use of quarantines and other regulatory practices. While it is probable that some nematodes not found in the United States would not become widespread here in any event, it is also probable that many would. This justifies the continued use of quarantines where there is separation by wide geographical barriers.

The occurrence of a species in two widely isolated areas indicates either (1) that it has been disseminated across wide geographical barriers, (2) that the distributions were once linked and have become separated by historical events, or (3) that there was parallel evolution. If physical barriers of large dimensions are effective deterrents to nematode dispersal, then one may ask, at what size and in what situations does a geographical force cease to be effective? The answer is not known, but it is sure to vary considerably with different situations. Although large bodies of water are effective barriers in many cases in geographical isolation, the importance of land barriers is more difficult to assess. Part of this is due to the fact that the United States has not been surveyed ecologically for nematodes. For example, the known distribution of *Merlinius joctus* in 1960 was Alaska and New Jersey (Thorne, 1949, 1961; Zuckerman and Coughlin, 1960). Since then it has been reported from a wheat field in North Dakota (Pepper, 1963); blueberries in Indiana, where it was thought to be introduced (Ferris and Ferris, 1967); blueberries in Michigan (Tjepkema et al., 1967); and a native wet prairie in Iowa (Schmitt and Norton, 1972). Further research may link the nematode's occurrence in the North Central Region to the cranberry bogs of the Northeast via the prairie peninsula of Indiana, Ohio, and Michigan and the moist habitats in forested areas.

As another example, *Tylenchorhynchus brevicaudatus* is known at 7500 feet in the Wasatch National Forest in Utah (Hopper, 1959), but it is not known to occur again until eastern Iowa, where it was found in a deep ravine (Norton et al., 1964). One would suppose that the migration route from Utah to eastern Iowa, or vice versa, would be along the deciduous forest outliers of the plains, and possibly future samplings will reveal this. It should be remembered, however, that the plains were not always the dry areas that they are today and might once have provided more favorable habitats than present climatic conditions permit. The examples of *M. joctus* and *T. brevicaudatus* have been given because these nematodes evidently have specialized habitats. By studying them, we might be able to learn more about dispersal than we can from nematodes having less specialized habitats.

Nematodes can be influenced by factors ranging from continental climates to those in the environment of the root hair. Some continental aspects will be mentioned here. Certain nematodes such as *Tylenchus davainei, Pratylenchus penetrans,* and *Xiphinema americanum* seem to be distributed generally throughout the United States and into Canada, although reports of the latter two nematodes are much more frequent from the eastern half of the continent. *Anguina agrostis* seems to be a problem only in the northern latitudes. Several nematodes are known mainly in the southeastern states, or at least in the warmer climates, as in California (Siddiqui et al., 1973); these include *Meloidogyne arenaria* and *Pratylenchus zeae.* Some species are tropical or subtropical, being found only as far north as the southern United States. These include *Radopholus similis* and *Rotylenchulus reniformis*, both of which have a wide host range suggesting that their northern limitation is the result of climatic conditions. Whereas many distribution patterns possibly reflect the interests of nematologists, all of the above-mentioned nematodes are well known and are conspicuous by their absence in areas where competent nematologists have worked.

Limitations in other directions are not so easy to deduce. *Nacobbus* spp. in the United States are known only in the central and western states, but it is not known whether this is because of geographical isolation, host, climatic, or soil factors. There are major differences in rainfall and edaphic patterns among soils of the eastern and western United States, and these probably are important at times in limiting establishment of certain nematodes.

Siddiqui et al. (1973) speculate that most species of pathogenic nematodes in California are introduced. This speculation is somewhat surprising to me, but there may be an element of truth in it considering the diversity of agriculture in that state and the diversity of nematodes that the authors list as being intercepted during a 20-year period. *Hemicycliophora arenaria* (McElroy et al., 1966) and *Meloidogyne hapla* (Raski, 1957) are believed to be native to the state, however, both being found on indigenous plants in land that was never cultivated. Doubtless there are other instances. The question is: How much escape from cultivation is there in the relatively short time that the United States has been under intensive cultivation? It is well known that large numbers of economically important nematodes have been introduced into the United States and some of these have become established. There are probably many more instances than we know. Unfortunately, not enough uncultivated native areas have been explored to determine accurately which nematodes are indigenous and which are not.

Even in the interior areas of the United States where native tracts of vegetation are found as islands in a sea of cultivation, it is probably safe to assume that most nematodes in these natural areas are indigenous, because where would they have come from? Native tracts doubtless have served as a source of infestation into cultivated lands for important parasites such as *Xiphinema americanum* and *Hoplolaimus galeatus*.

Oostenbrink (1966) presents some interesting information on the rate of infestation in the replacement of a brackish water nematode population with a "normal" soil population in reclaimed polders in Holland. Incidental numbers of Tylenchida were found in one polder in the second year after reclamation, and notable numbers were found by the seventh year. These included *Criconemoides parvum* and *Helicotylenchus* sp. Populations of plant-parasitic nematodes were as dense and complex after 10 to 20 years in reclaimed soil as in old soil.

Introduced Species

Several foreign nematodes have been introduced and have become well established in agricultural areas of North America. One, *Heterodera rostochiensis*, has remained localized, while another, *Anguina tritici*, has become more widespread. Both probably have been restricted by their narrow host range. Theories on the origin of *Heterodera glycines* are more speculative, and there are arguments supporting both its indigenousness as well as the possibility of its being introduced. Other nematodes sometimes considered as being introduced are *Ditylenchus dipsaci, Heterodera avenae, Meloidogyne arenaria, M. javanica,* and *Radopholus similis*. These nematodes provide interesting discussions on the effect of humans on nematode distribution, on the use and effectiveness of regulatory programs, and on economics. We will discuss only a few of the probably introduced species.

Anguina tritici.   Since wheat is an introduced species in the United States, probably *Anguina tritici* is also, especially in view of its narrow host range. In spite of its introduction and spread to many parts of the country, this nematode apparently has become established in only the more humid regions of the southeastern states. Although it has been reported from California (E. C. Johnson, 1909; Byars, 1919), New York (E. C. Johnson, 1909), and Texas (Taubenhaus and Ezekiel, 1933), and experimentally in Wisconsin (Leukel, 1924), as far as the writer is aware, *Anguina tritici* has not persisted in these places. Other than the Texas report, it has never been reported in the great wheat-producing states of the Midwest. The unrestricted shipment of grain and the poor seed technology in the 1920s and 1930s would have allowed ample opportunity for dissemination of the nematode. Although historical factors, possible varietal isolation, and the narrow host range must be considered, climatic factors seem the most plausible reasons for the restricted distribution of the nematode.

Heterodera rostochiensis. A nematode that was probably *Heterodera rostochiensis* was found in Germany in 1881, but it was not confirmed until 1913. The nematode was known in Scotland and England between 1913 and 1917, but damage that was probably due to this species was present about 1900. *Heterodera rostochiensis* was first found in the United States in 1941, but again a history of unthrifty potatoes preceded that by several years (Chitwood, 1951b). Much evidence supports the supposition that the nematode was introduced into Europe from South America, where many hosts of the nematode are indigenous, and then into the United States from Europe. It is possible that the nematode accompanied the potato which was introduced into Virginia in 1621 from Bermuda, where it had been taken from England, and into New Hampshire from Ireland in 1719. But since the pest is not known in these areas, perhaps the most likely introduction was by the Dutch at New Amsterdam, now New York City. A later introduction is also possible, as outbreaks in North America have been localized.

The main infestation of the nematode in the United States has been in Long Island, New York. More recent small findings near Syracuse, New York, and in New Jersey and Delaware are believed to have been spread by human beings. If the nematode were indigenous to the United States and could persist on weedy species of *Solanum*, one would expect it would be much more widespread than it is. It also attacks eggplant, which is a native of India and China, and the cultivated tomato, native to the Andes. Thus distribution patterns and host relationships strongly suggest that *H. rostochiensis* is not indigenous to the United States.

Heterodera glycines. The author is inclined to agree with the frequent speculation that *Heterodera glycines* was introduced into the United States, but there is increasing support to the theory that this nematode is indigenous. The fact that *H. glycines* is known only in certain soybean-growing areas and that soybeans are so widespread would indicate that the nematode was introduced and is spreading. The alternative is that it is indigenous and is resticted environmentally. The soybean is native to the Far East, where it has been cultivated for more than 5000 years, and the soybean cyst nematode is known to occur in Korea, Manchuria, and Japan. The soybean was introduced into the United States about 1800. Most introductions have been from Asia, but a few came from Europe. Approximately 83% of the soybeans grown in the United States about

1960 were produced in the north central states. If the soybean cyst nematode were indigenous to the United States, and if the northern climate was not restrictive, one would expect that the nematode would have been found more in the major soybean-producing areas of the north central states. The nematode is known to occur mostly in the South and Tennessee and Missouri, with its northern limit being central Illinois. The probability is that, if the many weed hosts are effective in the field, the nematode, if indigenous, would occur widely in the North. It is also interesting to note that *H. glycines* came to the fore in the lower Mississippi Valley as soybean was replacing cotton over large acreages. If the nematode was not introduced, this series of events would indicate that a low-level indigenous survival suddenly erupted when large acreages of highly susceptible germ plasm became available.

Heterodera trifolii.   *Heterodera trifolii* is one of the most widespread cyst nematodes in the United States. Whether this nematode was introduced or not, it is evident that the agricultural revolution in the United States has been favorable for this species. The nematode is common on white clover in turf and no doubt is disseminated widely through this vector. Many of the highly susceptible and widely cultivated hosts in the United States were introduced from Europe or Asia Minor in the seventeenth to nineteenth centuries. These include *Trifolium fragiferum, T. michelianum, T. pratense, T. repens, Vicia villosa, V. dasycarpon,* and *V. narboensis.* Thus there were good opportunities for introduction of the nematode. The alternative, as with *H. glycines,* is that the nematode is indigenous to the United States and survived in small numbers on noncultivated hosts.

NEMATODE CLUSTERING

As we progress from continental to regional to local distribution, we ultimately come to the field distribution and the plant itself. Nematodes are basically distributed uniformly, randomly, or in clusters (Figure 1.2). The most common pattern of spatial distribtuion of nematodes is the cluster, resulting in the well-known "pocket effect." Clusters are often not well defined and may be any arrangement in which nematodes may seem to be

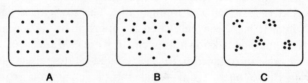

A   B   C

Figure 1.2.   Patterns of distribution. A, uniform; B, random; C, clustered.

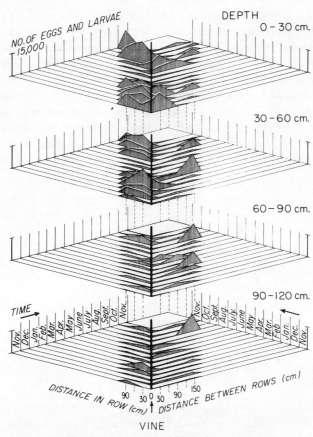

Figure 1.3. Spatial distribution of *Meloidogyne* spp. on grapevine roots. Points plotted are the average number of nematodes in 500 cc of soil from the upper and lower 15 cm of each depth interval. (From H. Ferris and McKenry, 1974.)

grouped together with intervening spaces separating them from other groups. Clustering of nematodes has caused unsolved problems in sampling. A species or several species may cluster in one place. Spatial clusters of nematodes result from spatial patterns of the present or previous hosts, morphology and distribution of the roots, almost imperceptible environmental changes, land management practices, fortuitous events of climate, inherent reproductive capacities, migratory ability, and other biological aspects of the nematode.

Nematodes probably are seldom, if ever, uniformly distributed. They may seem randomly distributed at times, but this may be a matter of delimiting areas. As Sneath and Sokal (1973) note, a distribution may be random on one scale and clustered on another. Barker and Nusbaum (1971) found that *Tylenchorhynchus claytoni* was fairly uniformly distributed in a field of barley following tobacco, but *Meloidogyne incognita* was not in a field of fescue following tobacco. Severe crop damage caused by a nematode can sometimes be fairly uniform in a field, but more frequently the nematode causes ragged and uneven stands or obliterates a crop in uniform to irregular spots. This pattern is largely a reflection of nematode population size. Differences in vegetation are major factors in many situations, and usually with vegetational changes in noncultivated areas there are concomitant edaphic changes that may be just as important. In noncultivated

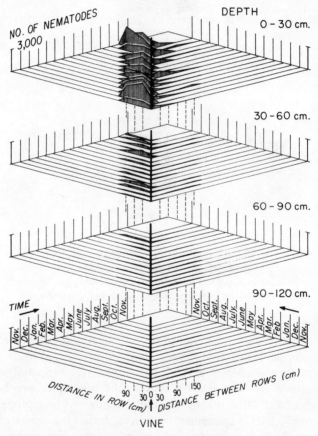

Figure 1.4. Spatial distribution of *Xiphinema americanum* on grapevine roots. Points plotted are the average number of nematodes in 500 cc of soil from the upper and lower 15 cm of each depth interval. (From H. Ferris and McKenry, 1974.)

prairies, Schmitt and Norton (1972) found that nematode distribution often was correlated with topography, which, of course, would include differences in vegetation. Yuen (1966) found that *Helicotylenchus paxilli, H. varicaudatus,* and *H. vulgaris* had their own centers of density in Broadbalk Wilderness in England, and that there was some correlation with vegetation. Similar results were found with other species of *Helicotylenchus* and with *Rotylenchus.* These findings are not surprising in view of ecological studies with other organisms.

Major differences in nematode distribution may occur in areas separated by only a few inches to a few feet in cultivated fields. Species of *Meloidogyne* and *Xiphinema americanum* had different spatial arrangements in vineyards, the latter being distributed mainly within the row and in the upper soil layers, while the former was more widely distributed throughout the profile and occurred frequently between as well as within the rows (Figures 1.3 and 1.4) (H. Ferris and McKenry, 1974). Thus there is something more than root distribution that can govern nematode occurrence. Changes in soil properties that seem small to people may loom large for the nematode. These differences may in part account for the seemingly erratic behavior of nematodes, and they will be discussed later.

CHAPTER 2

DISSEMINATION

Propagules are everywhere; the environment selects.

One can accept with assurance the ecological dictum that "propagules are everywhere; the environment selects," except of course, where geographical isolation has been caused by oceans, deserts, or other large barriers. With reference to nematodes, this dictum essentially states that if a sufficiently continuous food source is present, a species will occur in an area unless restricted by local environmental factors. For example, *Belonolaimus* species are restricted to the southern United States, not because they never were disseminated outside their present range but because they were unable to become established after they were disseminated. If one believes that *Belonolaimus* is restricted largely to the South because of lack of dissemination, he or she must then explain what is so peculiar about members of this genus that they cannot be disseminated when other nematodes can. Why is it that some nematodes have become established hundreds of kilometers north of the southernmost extension of the last Pleistocene ice in the past 10,000 to 14,000 years and others have not? Lack of dissemination is not the answer. Or have members of *Belonolaimus* evolved only recently? Are they still in the process of migration? In this chapter we will examine the evidence for active and passive nematode movement resulting in their local and long-distance spread.

PASSIVE DISPERSAL

There is little doubt that some aspects of dissemination are more important than others in nematode distribution. Active migration is nematode movement under its own locomotion, most of which occurs in the rhizosphere. Beyond the rhizosphere, however, the distance of active migration by a nematode is more controversial. There are reports that nematodes are capable of migrating several decimeters a year, but documentation is difficult. Much of the evidence on active migration in soil is circumstantial. Direct evidence is needed—such as recovering nematodes after tagging them with vital dyes or radioactive materials. Precautions should be taken, of course, to eliminate all possibility of passive transport. One may wonder if active migration for any distance is not greatly overshadowed by passive transport as an effective means of widespread dispersal. Evidence indicates that, unless carried passively, plant-parasitic nematodes usually spend most of their life in the vicinity of their birth, moving perhaps only a few centimeters a year. Unless this is proven to be untrue, we should be conservative in our estimates of nematodes' long-distance active migration. At the local plant level, it is difficult to distinguish between active migration and passive dispersal, but the former may assume major importance.

Dispersal by Water

Water is one of the important substances in life; without it, life would not long continue. Water not only allows the continuation of nematode species; it also disperses and even controls them. Although the greatest abundance of nematodes is found in marine habitats, we are interested primarily in terrestrial species. Marine nematodes are not the plant-parasitic forms, but their presence in such habitats indicates the suitability of water as a medium for many nematodes. Terrestrial nematodes are considered sometimes to be mainly aquatic in that they live either in sites having high humidity or in a film of moisture around their bodies. However, few plant nematodes can survive long in water where there is neither a food supply nor a rapid exchange of gases.

Water is a frequent means of passive dispersal. This may include surface runoff such as overland flow, streams, rivers, irrigation canals, percolation, interflow, and groundwater flow. Some methods are doubtless more important than others. Although one can conjecture much, information is sparse on nematode transport by water. Spread of nematodes by overland flow is assumed, but few data are available. *Heterodera rostochiensis* was stated to be spread 259 m by floodwaters in a single season (Chitwood, 1951b). *Dolichodorus heterocephalus* and *Hirschmanniella gracilis* are found along the shores of Lake Champlain (K. D. Fisher, 1968) and are doubtless carried to some extent in the water, as is probably true for any nematode in the littoral zone.

Dissemination of nematodes by irrigation water has been demonstrated conclusively by Faulkner and Bolander (1966) in the Columbia basin of eastern Washington. They found from 25 to over 200 nematodes per gallon (3.785 1) of irrigation water, and 10 to 20% of these were plant parasites. The greatest concentration of parasitic forms occurred in midseason and included *Ditylenchus, Paratylenchus, Pratylenchus, Tylenchorhynchus, Heterodera, Meloidogyne, Trichodorus,* and *Hemicycliophora.* An estimated 2×10^9 to 16×10^9 nematodes passed a given point per day (Figure 2.1). In a cross section of the canals, the nematodes did not concentrate in any position of the canal as long as the water was flowing (Figure 2.2); that is, there were no more nematodes per cubic foot of water in the faster than in the slower parts of the canal. The time of highest nematode counts varied by canal, being May in one and July in another. This depended partly on when waste was emptied into the canal. In later work (Faulkner and Bolander, 1970a), it was found that the primary source of plant-parasitic nematodes in the waterways was irrigation runoff into the irrigation system. Fumigated beds irrigated with canal water became heavily infested with plant-parasitic nematodes in three years, while those watered with well water did not (Faulkner and Bolander, 1970b).

Infiltration and percolation of water probably accounts for some downward nematode dispersal, but distance will vary with soil properties and precipitation. In temperate zones, percolation generally is greatest in the

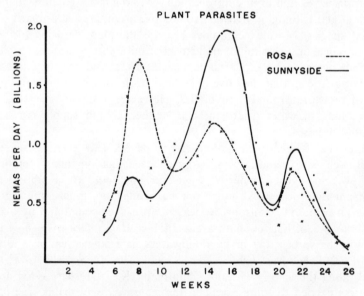

Figure 2.1. Estimated population of plant-parasitic nematodes per day passing the sample points over a 26-week period at two sites. (From Faulkner and Bolander, 1966, p. 598. Courtesy E. J. Brill Publishing Company.)

VELOCITY IN FEET PER SECOND (AVG.)

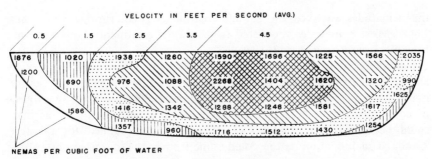

NEMAS PER CUBIC FOOT OF WATER

Figure 2.2.   Diagram of cross section of the Sunnyside Canal showing the number of nematodes collected per cubic foot (28.316 l) of water from each check point and the average velocity of water at these points. (From Faulkner and Bolander, 1966, p. 594.)

spring, decreasing with warmer temperatures and increased vegetative growth. It would seem that any downward nematode movement caused by percolation would follow similar patterns, although not to the depth that water occurs. The importance of percolation in carrying nematodes to the lower depths is questionable since most of the roots are in the top 30 cm, which is generally the area of greatest nematode activity. Movement of *Criconemoides curvatum* around carnations in observation boxes was slow and was mostly vertical rather than horizontal. The greatest vertical movement of the nematode was attributed to the downward percolation of water (Streu et al., 1961). It has also been shown that movement of *Radopholus similis* is aided by percolation (DuCharme, 1955). Circumstances vary, and the investigator should evaluate each instance individually.

Interflow is lateral underground movement of water where percolation water is forced laterally when it comes in contact with an impervious soil layer. We usually see it as seepage. Nematode movement in interflow is conceivable but has not been demonstrated. It is probably minor in overall dissemination of nematodes.

Groundwater flow is probably not a major source of nematode transport, since it is often deep and of long duration, during which time nematodes would most likely die. Few nematodes occur in groundwater, which usually is relatively pure because the soil filters out solids and microorganisms. When the groundwater table is near the soil surface, the probability of nematode inclusion and transport increases. Such an instance has been demonstrated by DuCharme (1955), whose finding of *Radopholus similis* in shallow wells in Florida is an indication that nematodes could be spread by subsoil water.

Spread by sewage sludge is a possible method of nematode transport that is due more to man's activities than to natural phenomena. *Heterodera schachtii* has been recovered from such sludge, but it is not known whether the nematodes survived the sewage treatments (Petherbridge and Jones, 1944). Most nematodes recovered from sewage treatment plants have not been plant parasites (Murad, 1970).

Nematodes may sometimes be collected from city drinking water by letting water from a faucet run gently through a fine mesh sieve for several hours. Mostly nonplant parasites will be found. Such water may be a source of contamination in tests involving soil that has been sterilized, as there could be an eventual buildup of nematodes not planned for the test. Although such contamination could come from other sources, nonsterilized water is a possibility.

Dispersal by Wind

Although microorganisms can be carried great distances by wind, there is little information on either the distance soil nematodes can be blown or the duration of their viability. Probably the greatest carry would come from bare soils; the more vegetation on a soil, the less wind erosion there would be, and therefore the less nematode transport. One might surmise that there would be more aerial transport when plants are in the early stages of growth than later when not only the larger plants would act as windbreaks but their roots would also partially anchor the soil.

Petherbridge and Jones (1944) reported that *Heterodera schachtii* is spread by wind, and Chitwood (1951b) found that cysts of *H. rostochiensis* can be blown 285 m and still contain viable contents (Table 2.1). White (1953) found that recovery of *H. rostochiensis* cysts declined sharply in collection traps that were placed higher in the air compared with traps near the ground (Table 2.2). Orr and Newton (1971) collected nematodes of 28 genera either from dust traps raised above the ground or from drift of wind-

Table 2.1. Wind dispersal of *Heterodera rostochiensis* cysts collected after one dust storm when much of the ground was covered with snow. Cysts and debris were collected in snow on top of uninfested land. Adapted from Chitwood (1951b)

Distance from heavily infested field (m)	Debris and soil collected (g)	Cysts (No.)	"Viable" (No.)
174	115	78	3
201	135	83	11
285	14	?	1
396	10	—	—

Table 2.2. Collection of cysts of *Heterodera rostochiensis* in 15 in.^{2} (97 cm^2) traps (March 28, 1953, Britain). Wind 4.8 km away was 25 knots. Modified from White (1953); reprinted from *Nature (Lond.)* 172:686

Height of trap (cm)	Soil (g)	Soil (cc)	Fully viable cysts (No.)	Semiviable cysts (No.)	Nonviable cysts (No.)
15–53	20,012	17,798	2,077	1,870	9,598
58–97	3,125	3,860	435	567	2,666
102–140	768	970	103	128	447

blown soil in Texas. Many nematodes were viable, and about half of the genera contained stylet-bearing forms, including *Criconemoides, Helicotylenchus, Meloidogyne, Pratylenchus, Merlinius,* and *Tylenchorhynchus.* Thus, wind probably is a more important method of nematode dissemination than has been documented. This importance increases as more land becomes extensively cultivated, with consequent exposure of soil.

Dissemination by Animals

Dissemination of plant-parasitic nematodes by animals must be considerable at times. Documentation is difficult, however, and little is know about distances covered. Since nematodes can pass through the digestive tract of animals and remain infectious in some instances (see Chapter 8), the effective distance of dissemination is governed by the mobility and speed of the vector and by the survival capacity of the nematode.

The best documentation that animals, namely insects, are important in the dissemination of nematodes injurious to plants is found in members of the Aphelenchoidea, especially those species that attack aerial parts of plants. Many Aphelenchoidea are well-known parasites of insects, but some species are economic plant parasites. It is these nematodes that are carried by insects that particularly interest us. There is strong evidence that *Rhadinaphelenchus cocophilus,* the cause of red ring of coconut palm, is carried by the palm weevil, *Rhynchophorus palmarum* (Blair and Darling, 1968; Hagley, 1963). Hagley (1963) reported that, although the main source of contamination appeared to be the frass in tunnels excavated by insect larvae, adults also could become contaminated. The nematode was also found internally after the insects fed on infected tissue. Longevity of the nematode on the body surface was from 2 to 6 days and as long as 10 days in the body cavity. The incidence of red ring was significantly reduced following application of an insecticide for control of the palm weevil. In another instance, *Bursaphelenchus lignicolus,* believed to be the cause of wilting of pines in Japan, was transmitted on the bodies of Cerambycid beetles, of which *Monochamus alternatus* seems to be the most important (Mamiya and Enda, 1972). It has also been established that Dipterous insects were the main means of dissemination of *Aphelenchoides composticola* in a commercial mushroom house (Haglund and Milne, 1973). An insect control program provided a practical solution to the nematode problem.

Dissemination by Foliage

Probably the best known active migration above ground is by *Aphelenchoides* spp. in which the nematodes move up the external plant parts and then invade the leaves. Wallace (1959b) speculated that migration of *A. ritzema-bosi* up chrysanthemum stems is possibly a negative geotrophic response in stationary water films but that this upward movement of nematodes is counteracted when water moves down the stem. The greatest mobility was in thick water films rather than in thin (thinner than the diameter of the nematode) and where there was a high concentration of epidermal hairs. Movement was by undulatory propulsion except in very thin layers. Invasion, which was through the stomates, occurred only when the water film was thin and the speed of the nematodes was slow. The geotaxis theory was not supported by migration of *A. besseyi*, but migration was favored by textured surfaces and an inverse water gradient (Adamo et al., 1976).

Marlatt (1970) has shown that *Aphelenchoides besseyi* can spread by contact from inflorescences of *Sporobolus poiretii* to *Ficus elastica* leaves. It was not ascertained whether the nematodes migrated or if wind, insects, or other factors helped them along. The results indicated that the nematodes moved from the soil to plants of *Sporobolus* and then to *Ficus* leaves, and that the nematodes did not move up the *Ficus* stems to the *Ficus* leaves. Theoretically, nematodes could move along any surface, providing there is adequate moisture and there are no inhibiting substances emanating from the plant. Any foliar inhibiting nematode, such as *Ditylenchus dipsaci* and *Aphelenchoides* spp., can be disseminated passively by leaf drop and wind-blown leaves.

Dissemination by Other Means

Nematodes can be disseminated passively in so many ways that it is difficult to estimate the importance of each. Any manner of carrying soil or plant material is a means for carrying nematodes. One has only to examine the records of intercepted pests at ports of entry to realize that plants and soil are nematode carriers. For the fiscal year ending June 30, 1968, 26 species of plant-parasitic nematodes, in a total of 226 interceptions from at least 31 different countries plus Hawaii, were recorded as being intercepted by the U.S. Department of Agriculture. Interceptions were made from plants, soil on plants, soil on shoes, and soil on automobiles and tractors (Girard, 1969). Farm machinery (Hagge, 1969), soil peds (Epps, 1969; Hagge, 1969), and even airplane tail skids (Thorne, 1934) may transport nematodes.

The predominant method of nematode dispersal at any one time will depend on a multitude of environmental factors, as well as on the inherent survival capabilities and the life cycle of the nematode. The important fact is that nematodes do become dispersed by many mechanisms. Understanding the principal methods, however, might weigh heavily in determining the feasibility or nonfeasibility of control, preventive, or other regulatory measures concerning nematodes.

MOVEMENT IN SOIL

The significance of active migration should not be underestimated at the plant level because, although roots move through the soil to the nematodes to a great extent, establishment of a nematode may depend on the nematode's ability to move short distances in the rhizosphere. Thus, soil factors such as texture, moisture, and temperature and their interrelationships play important roles in this minimal movement. These and other aspects will be discussed in Chapter 7.

The distances that nematodes migrate are not well known. Because it is often difficult to differentiate between active migration and passive movement, it is also difficult to ascertain the accuracy of distances that have been arrived at by deduction. Many experiments on migration are conducted under artificial conditions that essentially represent their own peculiar microclimates. Thus, the results might be biased one way or the other relative to migration under more natural conditions. Migration of nematodes on agar has been studied but often relative to attraction to roots or fungi or for thermal gradient investigations. Solid substrates probably favor nematode migration. It is well known that *Aphelenchoides* can migrate with apparent ease up the stems of chrysanthemums and other plants. Corn roots possibly facilitated movement of *Pratylenchus penetrans* in the soil as reported by Endo (1959). Tracks on agar are frequently random (Sandstedt et al., 1961), and movement can be rapid. A solid substrate such as agar can also facilitate diffusion of chemical substances in a manner not duplicated in soils. Comparing movement on solid substrates with that in soil labyrinths can easily lead to misinterpretations.

Horizontal Movement

There has been more experimentation with vertical than with horizontal movement. Perhaps this is partly because of our curiosity about the importance of vertical migration to deeper layers, often with the tacit assumption that nematodes are avoiding colder temperatures in the upper soil layers during winter. Cause and effect relationships should be clearly understood. Part of the greater interest in vertical migration may also be due to speculation that nematodes move to specific root zones because of nutritional aspects or chemical attractants.

Webster (1964) has given some insight about horizontal movement of *Ditylenchus dipsaci* in 46-cm diameter microplots planted with narcissus. He planted a total of 48 clean bulbs in concentric rings around a single infested bulb. Plots were sampled periodically, a separate microplot at each sampling time, and the number of nematodes recovered from each bulb was noted. The nematodes spread randomly outward from the center of infestation, causing new loci of infection in plants in adjacent rings. Many nematodes apparently migrated to the periphery of the microplots, passing host plants on the way. Thus it appears that if any plant exudate was produced it diffused only a short distance through the soil, or it had a short tenure in the soil, or the nematodes were not receptive. The author suggested that the spread of infestation was slower in the microplots than in the field, where there was extrinsic factors such as cultivation and water runoff. Webster postulated that the rate of spread of the nematode is contolled by (1) the net rate of increase in the nematode population, (2) the movement of the nematode through the soil, and (3) the external agencies causing dispersal of the nematodes. Later, Webster and Greet (1967) found that fallow plots were invaded more slowly by *D. dipsaci* than plots in which oats were planted. In the early stages of the experiment, *D. dipsaci* moved from nonfumigated fallow plots faster than from the oat plots. Thus it would appear that with small nematode populations the host plant is important in slowing the spread of the nematode. Cultivation across fumigated and nonfumigated plots did not increase the infestation by *D. dipsaci*. This is contrary to observations that circumstantially suggest that other nematodes are spread by cultivation.

Endo (1959) found no evidence of migration of *Pratylenchus zeae* after four months in fallow soils of three textures. In the presence of corn, however, migration was greatest in a sandy loam, intermediate in loam, and least in clay, Thus, plant roots were conducive to migration. Townshend and Webber (1971) found that *P. penetrans* moved more in soils of low than high bulk densities, but that optimum movement was only about 2 cm in seven days.

Vertical Distribution

It has long been said that most soil nematodes are in the top 30 cm of soil, and often mostly in the top 15 to 20 cm. This has been based upon many observations and it is doubtless true although exceptions occur. Probably more work has been done on the vertical distribution of migratory ecto- and semiendoparasites than with any other group of nematodes. Only a few records will be mentioned. Richter (1969) found that most *Tylenchorhynchus* spp. were in the 0 to 20 cm depth in a sandy soil, many fewer being found in the 20 to 40 cm or the 40 to 60 cm depths. Several workers in the northern Midwest (Griffin and Darling, 1964; Norton, 1963; Schmitt, 1973b) reported that the greatest numbers of *Xiphinema americanum* occurred in the upper layers of the soil, and Cohn (1969) recorded that seven species of *Xiphinema* and two of *Longidorus* in Israel occurred in the 0 to 30 cm layer and that they decreased with increasing depth. Most *Ditylenchus dipsaci* were in the top 15 cm of an organic soil in southern New York in late summer and early fall following onions (Lewis and Mai, 1960). Using oats in a clay-loam soil and studying the top 20 cm, Wallace (1962) found that the greatest numbers of *D. dipsaci* were near the surface and decreased with depth during rainy periods. Nematode densities were greatest in the 0 to 10 cm depth, and generally declined to 30 to 60 cm depths, the deepest studied in 6-year-old alfalfa in silty-loam soil (Tseng et al., 1968). Species of *Ditylenchus*, such as *D. radicicola*, that parasitize roots may have generally deeper soil distributions, but information is not available.

As usual there are exceptions to the rule, and conditions exist where nematodes are less numerous in the upper soil layers than in deeper layers. Although Richter (1969) found that *Trichodorus* spp. had a somewhat even distribution in a sandy soil, greater concentrations occurred 20 to 40 cm deep. More *T. viruliferus* occurred in the 10 to 40 cm range, while *T. pachydermus* was present mainly between 30 and 60 cm. *Helicotylenchus vulgaris* had different vertical distributions in similar soil types but dissimilar vegetations, while *H. vulgaris* and *Rotylenchus pumilus* had different vertical distribution patterns in the same habitat (Yuen, 1966). Zuckerman et al. (1964) found that the greatest concentration of *Hemicycliophora zuckermani* in old cranberry bogs was in the root zone 7.6 cm deep, while in young bogs of 10 years production the greatest numbers were 10 to 15 cm deep. In the young bog, the top 2 to 3 in. was almost pure sand with a low water-holding capacity.

It has been demonstrated that nematode distribution often is correlated with root distribution, not a surprising observation since the obligate nature of most plant parasites dictates a living food source. Wallace and Greet

(1964) found that, in a meadow of *Phleum pratense* and *Festuca arvensis, Tylenchorhynchus icarus* was most concentrated at about 5 cm depth, which was also the region of greatest root concentration. Few nematodes were found below 24 cm, which was the boundary between the topsoil and subsoil. Mean numbers of *X. americanum* decreased with increased soil depth as the fibrous root weight decreased in profile depth for lilacs in two different soils (Schmitt, 1973b). Also, Krebill et al. (1967) made a positive correlation of maximum *X. americanum* numbers with the number of lateral roots of *Pinus banksiana*.

Even though susceptible roots exist in the top layer of the soil, other factors may preclude nematodes occurring in greater numbers than in deeper layers. The lowest percentage recovery of *Radopholus similis* from citrus in Florida was in the top 30 cm of soil, but numbers of this nematode increased with increasing depth (DuCharme, 1967). Suit et al. (1953) found *R. similis* at all soil depths down to 2.4 m, but most were between 0.3 and 1.5 m. The nematode also was dissected from roots at a depth of 3 to 3.7 m. Fibrous root development in the top 30 cm of soil did not differ markedly in citrus groves infested and uninfested with *R. similis*. This nematode evidently is able to survive adverse conditions in these fibrous roots, but its development is limited much of the year because of high soil temperatures in the top 30 cm of soil. About 25 to 30% of the fibrous roots were lost at the 25 to 76 cm level, but below 76 cm up to 90% of the fibrous roots died within a few weeks after being formed (Ford, 1953).

Thus it seems that most of the nematodes studied are in the upper layers of the soil and where there are the most roots, but exceptions occur (H. Ferris and McKenry, 1974) (Figures 1.3 and 1.4). As one would expect, root morphology and distribution have much to do with the location of nematodes. But, as Brodie (1976b) found with three species of nematodes around soybeans, other factors can be important.

Vertical Migration

The nature and amount of vertical migration have been debated much. Although there seems to be no consistent pattern, it is becoming clearer that active vertical migration does occur even though it evidently is not a universal phenomenon.

Lewis and Mai (1960) found more *Ditylenchus dipsaci* in two onion fields in New York in the 15 to 38 cm layer than in the 0 to 15 cm layer in the winter, but the reverse was true in both fields in the fall and in one field in the spring. They interpreted this as resulting from partial downward or upward migration. Wallace (1962) found that surface populations of *D. dipsaci* declined in the absence of rainfall, but his field data did not support a migration concept. He found, however, that some upward migration occurred in soil tubes and that this was accentuated by a host plant, a phenomenon also recorded for *X. americanum* by Schmitt (1973b). Streu et al. (1961) found that movement of *Criconemoides curvatum* was slow, and that most movement in observation boxes was vertical rather than horizontal. They concluded that water percolation was a major factor in the greater vertical than horizontal distribution. Using *Tylenchulus semipenetrans*, Baines (1974) likewise found little nematode movement downward in a pot and concluded that there was little migration.

While studying *Xiphinema americanum* in alfalfa plots in Iowa, Norton (1963) found no consistent evidence for migration from one level to another, although there was some indication that migration might be occurring in one test—that is, there were greater numbers of *X. americanum* in the upper than the lower level during the summer, but the reverse was true in the early spring and in the fall. It was speculated that this might not be migration per se and might result from differential activity due to temperature changes at the two depths. There was no such "migration" phenomenon at a second station. There was no evidence for vertical migration of *Longidorus elongatus* (Wyss, 1970a), *Trichodorus viruliferus, T. pachydermus* (Richter, 1969), and *T. christiei* (Hoff and Mai, 1964).

Some of the best evidence for vertical migration was supplied by Rössner (1970, 1972), who, by using sections of soil-filled plastic tubes, was able to follow migration of several nematodes in an environment resembling that out of doors. *Trichodorus pachydermus, T. sparsus,* and *Longidorus* sp. moved quickly to deeper soils, some *Trichodorus* reaching a depth of 200 cm in 100 days. *Rotylenchus robustus* and *Helicotylenchus pseudorobustus* scarcely moved downward even when the host plants were well rooted. *Pratylenchus penetrans* and *Rotylenchus robustus* favored depths between 0 and 25 cm. Different stages of *P. penetrans* colonized different layers to different degrees. Planting with barley and red clover generally led to deeper soil penetration with all species tested. *Trichodorus* species lived in deeper moisture layers and also, by increased reproduction in autumn, could survive independent outside conditions.

There has been much speculation concerning temperature as a possible stimulus for vertical migration. We have seen that nematodes do have different levels of "preference" and that there is undoubtedly active migration under some circumstances. It is often stated, usually without much evidence, that nematodes move downward in winter "to get away from the cold," but assuming that there is considerable vertical migration—at least more than we know about—how much is due to temperature changes and how much to other factors? Let us look at some of the evidence on how nematodes respond to thermal gradients.

Rode (1969) found that movements of *Heterodera rostochiensis* on an agar substrate decreased above 30 C, and all movement stopped at 42 C. The nematodes were sensitive to small temperature changes, there being a distinct orientation toward the temperature preferendum with gradients of only 0.016 to 0.02 C/cm from some initial temperatures. In other work, it was noted that on agar *Ditylenchus dipsaci, Pratylenchus penetrans,* and *Tylenchorhynchus claytoni* responded to a temperature gradient of 0.033 C 4 cm from a heat source, which could be heating wires, infrared radiation, or germinating alfalfa seeds (El-Sherif and Mai, 1969). *P. penetrans* moved toward the heat source from a 1-cm distance in 10 min. The nematodes dispersed when the heat was moved but reassembled when the heat was restored. Germinating seeds, in the absence of CO_2, attracted *P. penetrans. Trichodorus christiei* and *Xiphinema americanum* did not respond to the heat treatments. *Ditylenchus dipsaci* had a temperature preferendum in sand of 10 C in a temperature gradient of 2 to 30 C (Wallace, 1961), but Croll (1967) found that *D. dipsaci* became acclimatized to a storage temperature. When nematodes were stored at 10, 20, or 30 C and then subjected to a heat gradient, they accumulated and showed the greatest activity at the storage temperature. It is evident that some nematodes can respond to temperature gradients, but those tested in soil have been tested under relatively short distances and under artificial conditions. *Trichodorus christiei* did not respond to heat, but Rössner found that some species of the genus moved in soil. Is movement as elucidated by Rössner caused by factors other than temperature, or are there differences among species within a genus in their reaction to temperatures, or are the experimental conditions so diverse that comparisons and speculations are unwarranted? Natural diurnal temperature fluctuations in the soil would certainly override small gradations as measured in laboratory experiments. Nevertheless, such experiments are worthwhile, but much has to be done with many species under many conditions before the importance of temperature in vertical migration can be evaluated with confidence.

ATTRACTION

There has been ample confirmation of Linford's (1939) observation that certain nematodes are attracted to roots. This phenomenon indicates that there is some nematode movement in the rhizosphere, but it also implies that not all movement is random. Although Linford was not the first to propose that nematodes are attracted to roots, his observations acted as a catalyst for much experimentation, sometimes with seemingly contradictory results. But, as Wallace (1963) pointed out, these differences might reflect limitations in techniques. Without getting deeply into the physiological aspects of attraction, we can accept the observations that attraction does occur in some instances, but that qualitative and quantitative differences occur (Barrows, 1939; Blake, 1962; Doncaster, 1953; Shepherd, 1959). It is well known that some nematodes, especially *Meloidogyne* spp., are attracted to an area just behind the root tip, while others, such as *Pratylenchus* spp., sometimes are attracted to the root tip and sometimes farther back. When Linford (1939) mixed populations of *Meloidogyne* sp. and *P. pratensis*, the nematodes separated into two distinct zones.

That the area of the root tip is attractive is not very surprising when one considers that it is a region of high metabolic activity from which numerous substances diffuse, some of which might act as attractants, some as repellants, and some neither. This has been the source of much investigation. Wieser (1955) found that the apical 2 mm of excised tomato root tips were repellent to *Meloidogyne hapla*, the next 6 mm (zone of elongation) were attractive, while the piliferous zones were neutral. Later work showed that excised roots of soybeans and eggplants exhibited no distinct areas of attraction (Wieser, 1956). Viglierchio (1961a) reported that *Meloidogyne hapla* was attracted to rye in soil but repulsed in sand. This phenomenon did not exist for all combinations of nematodes and plants; many times there was no difference in attraction in sand compared with soil.

Several workers (Blake, 1962; Johnson and Viglierchio, 1961; Lownsbery and Viglierchio, 1960; Viglierchio, 1961a) have demonstrated that compounds emanating from the rhizosphere can diffuse through a dialysis membrane and still be attractive. Generally these workers used a viscose-regenerated cellulose membrane with an average pore size of 24 Å; therefore, the attractant must be of low molecular weight. Several gases in the soil could penetrate this membrane. These include CO_2 (3.2 Å), (2.9 Å), N_2 (3.1 Å), and H_2 (2.7 Å). Sometimes, if bacteria accompanying the seedlings were eliminated, so was the attraction, an indication that some of the attractants might originate from organisms associated closely with the roots.

Several chemicals that attract nematodes under some conditions are known to emanate from roots. These include CO_2 (Klingler, 1961), giberellic and glutamic acids (Bird, 1959, 1962), and tyrosine (Oteifa and Elgindi, 1961). The latter authors found that tyrosine was attractive but that other amino acides were repellant or neutral. Bacteria attract juveniles of *Heterodera schachtii* (Bergman and van Duuren, 1959), and *Neotylenchus linfordi* was attracted to filtrates or colonies of several fungi (Klink et al., 1970). Of specific gases that have been considered to be attractive to nematodes, CO_2 has been among the foremost studied. Bird (1962) emphasized the importance of CO_2 as a primary attractant, but he also showed that galled tissue attracted twice as many *Meloidogyne javanica* juveniles as nongalled tissue, even though his studies found no significant difference in CO_2 production by either kind of tissue.

An interesting theory was proposed by Rohde (1960), who studied the influence of CO_2 on nematode respiration. Respiration rates decreased constantly from time of removal from the host plant until death occurred. Respiration rates of certain nematodes were higher in air than in vessels without CO_2 or with higher CO_2 levels than those in the air. Rohde speculated that CO_2 from respiring roots acts as an orthokinetic stimulus that decreases activity and prevents nematodes from leaving the root area.

Klingler (1963) conducted a series of experiments using nematodes in agar and artificial and biological CO_2 gradients, the latter being from germinating seeds. It was found by gas chromatography and electroconductometric samples taken at varying distances from CO_2 sources that the minimum concentration difference necessary for a directed orientation of *Ditylenchus dipsaci* was usually 0.08 to 0.15% CO_2 per cm. It was suggested that these directed movements may also occur under natural soil conditions up to 1 to 2 cm from the CO_2 source. Klingler (1965) also investigated the possibility of a descending O_2 gradient instead of an ascending CO_2 gradient being responsible for his 1963 results. Using N_2 or argon to diminish the oxygen, he found that *Ditylenchus dipsaci* did not exhibit any directed movement in a descending O_2 gradient but did in an ascending CO_2 gradient. Johnson and Viglierchio (1961) used an aluminum channel coated with asphaltum and inscribed at 1-cm intervals. Nematodes were evenly distributed in sand saturated with water, and a dialysis tube saturated with oxygen or carbon dioxide was positioned in the center of the channel. *Ditylenchus dipsaci, Meloidogyne hapla, M. javanica,* and *Heterodera schachtii* accumulated around a CO_2 source, but *Heterodera schachtii* and *Ditylenchus dipsaci* also accumulated around and O_2 source.

While it is not the purpose here to go into the physiological aspects, this is an important area, and the reader is referred to reviews by Green (1971) and Klingler (1965). Of more concern to us is that such movement can occur and the extent to which it can occur. Unfortunately, we know little about the latter, but indications are that it is generally not more than a few centimeters. Here again, most work has been performed under highly artificial conditions, and wide latitude should be allowed in interpretation. Luc (1961) reported that millet influenced nematode movement 40 cm away, but most workers report much shorter distances. The production and degration of attractants doubtless vary and depend on the associated plant, soil texture, moisture, and related soil factors that may affect diffusion patterns. Therefore, generalizations often have to be tenuous.

HABITATS

Similar conditions support similar or equivalent
communities, and conversely, similar communities are
evidence of similar conditions.

Different environmental factors modify the processes of all
species, and, therefore, their form and structure.

Cobb's (1915) statement on the ubiquitousness of nematodes is well known, and Winslow (1960) called attention to Bastian's (1865) less known statement written a half century before Cobb. Bastian wrote:

> I am inclined to believe that these free nematodes will be found to
> constitute one of the most widely diffused and abundant groups in the
> whole animal kingdom, rivaling, in the first respect at least, the almost
> ubiquitous Diatomaceae. . . . Thus, beginning with the land and
> freshwater species, I have found them in all specimens of soil examined,
> in moss, various species of lichen, about the roots of fungi, also the roots
> of grasses, and between the sheaths of their leaves, amongst the mud of
> ponds and rivers, on the freshwater Algae, amidst decaying liverworts
> and mosses, and on submerged aquatic plants.

Bastian's and Cobb's statements could almost apply to the plant-parasitic forms. Wherever plants become established, parasites are sure to follow. Nematodes are known to feed on plants from the Arctic to the tropics, from deserts to swamps, and from mountain tops to valleys. The plant damage they cause may range from little to complete destruction of the host. The varied hosts and habitats support a variety of nematodes that have evolved in ecological niches dictated by the biological and physical-chemical environment. Most plant-parasitic nematodes are obligate parasites, and their successful establishment obviously requires a continuing living food source, a favorable physical and chemical environment, and a biological habitat that is not limiting. Samples from most vegetational habitats indicate that nematodes have been eminently successful, but it will be noted quickly that some species are more successful than others. This is natural since all species do not possess the same ecological amplitude, including food substrate and competitive ability. Some nematodes are recovered commonly in large numbers, others are found consistently in small numbers, and others occur sporadically. One species may occur frequently in the presence of another, whereas other species rarely if ever cohabit. This is not happenstance; it is the result of evolutionary processes.

A nematode species sometimes can occur in seemingly diverse environments. *Sphaeronema californicum,* for example, was found as an endoparasite of *Umbellularia californica* near sea level on the shore of Tomales Bay in California and as a parasite of *Arctostaphylos* at 1676 m (Raski and Sher, 1952). Although detailed information was not provided, this difference in altitude should not necessarily be construed as evidence that the microclimatic habitats were dissimilar.

The food source is the most important factor in maintaining a population of plant-parasitic nematodes. Without it, a population could never originate, let alone continue to survive. This dependency on a living food source and various degrees of parasitism delimit many nematodes from otherwise favorable environments. Thus the narrower the host range and the more restricted the host is geographically, the more restricted is the nematode if its existence is based on host availability alone. Climatic and edaphic conditions and competitive and antagonistic biota may further limit a nematode's distribution and prevalence. The most topographically uniform areas are usually the first to be cultivated, leaving the more diverse habitats intact. The diverse vegetation in these undisturbed habitats contributes greatly to the diversity of the nematode fauna. Topography that appears uniform to us may contain sufficient diversity in drainage and soil properties to affect nematode patterns both qualitatively and quantitatively.

HABITAT TYPES

Our association of nematodes with different habitats is as old as the observations of nematodes themselves. Examination of the nematode lists of early investigators reveals that tylenchids were a minor part of the recorded fauna. Whether this was due to scarcity of these forms in the habitats sampled or to the investigator's lack of interest is not known. Present evidence indicates, however, that it was probably both, along with the fact extraction methods were not so sophisticated as those of today. As the economic role of nematodes in agriculture became clearer, a greater emphasis on investigation of the plant-parasitic forms naturally followed. Although nematodes are of demonstrated importance to economic crops, it is surprising that noncultivated areas have not been investigated more, since it is from these areas that many of our pests in cultivated fields were derived. Such studies could expand our knowledge of agriculturally economic nematodes, the changes brought about by cultivation, and the importance of nematodes in the natural ecosystem.

Establishment of a plant-parasitic nematode can occur no faster than the establishment of its host, and this is slow in native areas compared with cultivated ones. The drastic and widespread changes from a heterogeneous to a relatively homogeneous food source brought about by cultivation permits a rapid dissemination and establishment of nematodes favored by such an environment. One has only to examine lists of nematodes found in woodlands (S. R. Johnson et al., 1972; Springer, 1964) and prairies or range land (Orr and Dickerson, 1966; Schmitt and Norton, 1972; Thorne and Malek, 1968) to find many plant parasites that are not commonly found in lists of nematodes from farm soils in the area. The incidence and prevalence of many nematodes such as *Pratylenchus* are less in native habitats than in the subsequently cultivated areas. It is likely that these either had restricted food supplies or were poor competitors prior to cultivation but were favored by agronomic practices.

Habitat and Niche

According to most ecologists, niche and habitat are not synonymous [see Boughey (1973) and Pielou (1974) for discussion]. The habitat is sometimes referred to as the "address" of an organism—that is, a physical location where the organism will be found. But it tells nothing of the many parameters—biological, physical, and chemical—that allow an organism to occupy, survive, live, and reproduce in a particular space. The niche is an abstract concept of the the totality of environmental factors that allow an organism to function, sometimes with varying degrees of success. It is a fundamental set of characteristics that make evolution and productivity possible for the organism.

The niche is bounded by tolerance limits of all biological, chemical, and physical factors for a given organism. An organism near but within the tolerance limit of one parameter may barely survive, but once the outer tolerance limit is reached the organism cannot persist even if all other parameters are favorable. A fundamental niche is the totality of all parameters that would allow an organism to function were it not for competition with other organisms. An ecological or realized niche is one that does occur. A spatial niche is one that has not yet been occupied by an organism and is thus of little concern to ecologists since it is not functional. There is nothing to evolve and nothing to produce.

With nematodes as with other organisms, marginal distribution patterns frequently indicate marginal niches, and the center of activity may be some distance away. As a geographical example, *Pratylenchus zeae* is found commonly in the southeastern United States but is not known to occur in the colder regions of the North. Assuming, and we have little data, that the nematode is restricted northward by low temperatures, and assuming that the germ plasms of the northern cultivars are susceptible, then the marginal populations and infrequent occurrences found in southern Indiana and other states bordering the nematode's known distribution range probably reflect marginal niches. As another example, individuals of *Pratylenchus* and *Meloidogyne* may inhabit the same root, but their microhabitats and niches are not the same. Colonization by each of these nematodes is restricted to separate tissues within the root even though the nematodes' niches may be only a few cells apart.

A problem in nematode ecology is that many habitats are ephemeral due to either root destruction by indigenous biological agents or human cultural practices; some nematode populations do not have time to develop to maturity. Even though the fundamental niche is present, it does not become fully realized. This may be one of the basic reasons for major qualitative and quantitative differences in nematode fauna between native and cultivated systems.

It is evident that the totality of environmental factors creates niches whereby some nematodes increase to large populations that may exist for long periods. The same totality of environmental factors may provide a niche for other nematodes that persist in small densities. A change in the totality, however, may allow these latter nematodes to thrive in the new niches.

Nematodes in Prairies

Grasslands provide one of our foremost sources of energy, either directly or indirectly, but the occurrence and bionomics of nematodes in them have scarcely been studied. The grasslands of mid-America comprises a vast expanse extending from Alberta to Texas, from the tall-grass prairies of Iowa and Missouri through the mid-grass prairie and short-grass plains to the Rocky Mountains. The Prairie Peninsula extends eastward into the deciduous forests of Ohio and Michigan (Transeau, 1935). The diversity of the prairies doubtless accounts at least in part for the diverse nematode fauna therein. There are numerous accounts of isolated occurrences of nematodes in rangelands west of the Rocky Mountains, but there have been few studies of a given area, local or general.

Approximately 66 species of plant-parasitic nematodes of native sod were reported in five preliminary studies in the mid-continent (Table 3.1). Not included are such groups as the Tylenchinae, Psilenchinae, Neotylenchidae, and many of the Aphelenchoidea whose members are known or suspected to feed on fungi or whose parasitism has not been demonstrated. Bias in the table results from limited descriptions of taxa available to earlier workers, restrictions on taxonomic treatments, nonuniform extraction methods, varying interests of the investigators, and omission of isolated reports. In any event, the prairie nematode fauna, taxonomically, is imperfectly understood.

Heterodera are mentioned in several reports, but species were not identified in most instances. *Heterodera longicolla* was found in large numbers on the roots of buffalo-grass in Kansas (Golden and Dickerson, 1973). *Heterodera punctata* was found as a parasite of wheat in Saskatchewan, and Thorne (1928) speculated that it was probably native since the fields had been in cultivation only a short time. The nematode was also found in prairie sod in Alberta (Baker, 1957), lending support to Thorne's contention. Horne and Thames (1966), who found this nematode on *Poa annua* in a remote Texas canyon, suggested that the nematode might be indigenous or at least not recently introduced. There are several reports of the nematode occurring in wheat fields on the plains, and it appears likely that the source of infestation was native prairie. Cyst nematodes probably are much more numerous in the plains and prairies than we realize. The same can be said for *Anguina*. Many are parasites of grasses. *Anguina agropyronifloris* is known in seed lots of western wheatgrass from Kansas, Montana, and South Dakota (Norton, 1965a). A nematode that I presume to be the same was found in a native stand of western wheatgrass in Montana (Collins, 1966). A species believed to be a *A. graminophila* was found on leaves of *Calamogrostis canadensis* (Goto and Gibler, 1951), a plant that is common in wet prairies and marshes and is sometimes harvested as marsh hay. *Anguina calamagrostis* later was described from specimens from leaf galls on *C. canadensis* in Quebec and Ontario (Wu, 1967).

It is apparent from Table 3.1 that there is a diverse nematode fauna in the prairies and that many species that occur in these prairie remnants could have been a source of infestation for cultivated crops. It is unlikely that many nematode species were introduced into the grasslands from cultivation, at least in mid-America, because from where would they come?

Although species of *Pratylenchus* are not abundant in lists of nematodes from grasslands, they probably are more common than is known. Gotoh (1970) listed seven species of *Pratylenchus*, most of which came from

Table 3.1. Plant-parasitic nematodes found in some native prairies in the United States

Nematode	Location	Nematode	Location
Aorolaimus torpidus	I	*M. tessellatus*	I, K
Aphelenchoides fragariae	K	*Nagelus aberrans*	NGP
Criconema decalineatun	—	*Paratylenchus microdorus*	I, O
C. hungaricum	—	*P. nanus*	I
Criconemoides basili	O	*P. pesticus*	NGP
C. curvatum	I, O	*P. projectus*	—
C. discus	I, NGP	*P. vexans*	NGP
C. incrassatus	—	*Pratylenchus agilis*	NGP
C. macrodorum	O	*P. coffeae*	K
C. permistus	—	*P. penetrans*	K
C. pseudosolivagum	I, O	*P. scribneri*	NGP
C. vernus	O	*Quinisulcius acutus*	NGP
C. xenoplax	I, K, O	*Radopholus neosimilis*	K
Discocriconemella inaratus	—	*Rotylenchus buxophilus*	K
Ditylenchus dipsaci	K	*R. robustus*	K
D. myceliophagus	I, K	*Trichodorus christiei*	K
Geocenamus tenuidens	NGP	*T. proximus*	K

Species	Location
Gracilacus aciculus	I
Helicotylenchus californicus	NGP
H. digonicus	I, K, O
H. dihystera	I
H. exallus	I, O
H. hydrophilus	I
H. labiodiscinus	I, O
H. leiocephalus	I
H. platyurus	I, K, O
H. pseudorobustus	I, K, O
Hoplolaimus galeatus	I, K, NGP, O
H. tylenchiformis	K
Merlinius joctus	I
M. leptus	I, O
M. lineatus	K
M. stegus	NGP
Trophurus imperialis	NGP
Tylenchorhynchus cylindricus	K
T. dubius	K
T. kegenicus	K
T. lamelliferus	K
T. martini	I
T. maximus	I, K, NGP, O
T. nudus	I, K, NGP, O
T. pachys	NGP
T. parvus	K
T. robustus	NGP
T. silvaticus	I, K
T. striatus	K
Xiphinema americanum	I, K, NGP, O
X. chambersi	I
X. index	K

I = Iowa (Hoffmann and Norton, 1973; Schmitt and Norton, 1972).

K = Kansas (Orr and Dickerson, 1966).

NGP = Northern Great Plains (Thorne, 1974; Thorne and Malek, 1968).

O = Ohio (R. M. Riedel and Norton, unpublished).

natural grasslands, found in Japanese noncultivated soils. Further studies are needed to elucidate and amplify our knowledge of distribution patterns, diversity, and associations existing within prairies. Only meager beginnings have been made. For example, Schmitt and Norton (1972) found, not surprisingly, that different patterns of clustering occurred in Iowa prairies. *Xiphinema americanum* clustered with the Mononchidae; these were in greatest abundance on the well-drained slopes. *Tylenchorhynchus nudus, Helicotylenchus pseudorobustus,* and small numbers of *X. americanum* were found on the summit and high shoulder sites. *Merlinius joctus, H. hydrophilus,* and *X. chambersi* were found only in potholes. Site habitat information for the first three nematodes seems to agree with observations in cultivated soils, although more statistical comparisons are desired.

Nematodes in Woodlands

Although many observations of nematode occurrences and damage to trees are recorded for nurseries, and to a lesser extent in plantations, there is comparatively little information on nematodes in natural woodlands and forests. Of the 237 references to plant-parasitic nematodes associated with forest trees, as collated by Ruehle (1967), most seem to be isolated reports of cursory surveys of incidental observations.

Parasitism of nematodes on trees has been demonstrated, and large increases in nematode numbers have been obtained in greenhouse tests (Riffle, 1970b, 1972; Ruehle, 1966, 1968a, 1969a, 1969b, 1971, 1972a, 1972b; Sutherland, 1967a, 1967b). Interpretations relative to field conditions are difficult because of differences in root distribution and density, biological antagonisms, and edaphic factors. Such increases in greenhouse tests indicate the nematode's potential and afford opportunity for evaluation of limiting factors in the field. A few studies of nematode occurrences in the United States have been made (Table 3.2). Nematodes listed as occurring in woodlands might not feed on true woody or herbaceous plants of a natural forest but might feed on weedy grasses present and therefore may be only incidental to woodlands. For example, Riffle (1968) suspected that the large population of *Helicotylenchus* in ponderosa pine stands in New Mexico was due to grass in the plots. Nevertheless, comparison of reported occurrences of nematodes in prairies and woodlands increases our knowledge of different distribution patterns between major vegetational formations. Of the 66 species listed as occuring in the prairies (Table 3.1) and the 69 reported from the woodlands (Table 3.2), only 22 are colisted. It can be expected that more species will be colisted as more data accumulate. It also can be expected that when frequencies and abundances are better known, qualitative as well as quantitative differences between and within major vegetational formations will become more apparent.

Root-knot, cyst, and other nematodes, often not recovered by routine soil processing, have been found as parasites of forest trees on several occasions. These include *Meloidogyne ovalis* on deciduous trees in Wisconsin (Riffle, 1963), *Heterodera betulae* on birch in Arkansas (Hirschmann and Riggs, 1969), *Meloidodera floridensis* on pines in North Carolina (Ruehle, 1962), and *Nacobbodera chitwoodi* on conifers in Oregon (Golden and Jensen, 1974; Zak and Jensen, 1975). These types are probably more common in natural stands than is now documented.

One would expect certain nematodes to be found in certain habitats, and a few observations in woodlands substantiate this. Using ordination techniques, for example, S. R. Johnson et al. (1973) concluded that *Helicotylenchus platyurus* was a mesic species, frequently being dominant in, but not restricted to, swamp forest communities in Indiana. A similar conclusion was reached in Iowa where the nematode was more mesic than *Xiphinema americanum, H. platyurus* being more common in bottomlands where *X. americanum* was infrequent (Norton and Hoffmann, 1974). Neither study indicated whether moisture or other factors were primarily responsible for the distribution patterns. There are also only fragmentary data on frequency and abundance of nematodes in specific forest types and microenvironments. It would be expected, however, that major differences exist. For example, *Helicotylenchus platyurus* and *Xiphinema americanum* are found only rarely in the hemlock-hardwood and boreal forests in the Adirondack Mountains of New York, and in Vermont, New Hampshire, and Maine (Norton and Hoffmann, 1974). These species are common, however, in the deciduous forests of Iowa (Norton and Hoffmann, 1974) and Indiana (S. R. Johnson et al., 1973). Differences were found even in two closely associated formations. *Bakernema inaequale, Criconema menzeli, C. octangulare, C. proclivis, Criconemoides petasus, C. sphagni,* and *C. xenoplax* occurred in the hemlock-hardwood communities, while only *C. menzeli* and *C. sphagni* were found in the boreal forests of the northeastern mountains (Hoffmann and Norton, 1976).

Desert Fauna

Certainly there must be a desert nematode fauna, but little has been done to characterize it taxonomically or ecologically. Thorne's collections in the western United States contain many specimens from desert habitats, and doubtless a collation of this information would be a major contribution. Freckman et al. (1974) identified only five tylenchids around desert shrubs

Table 3.2. Plant-parasitic nematodes found in some native woodlands in the United States

Nematode	Location	Nematode	Location
Bakernema inaequale	BH, NJ	*Hemicriconemoides chitwoodi*	SHC
Criconema cobbi	In	*Hemicycliophora gigas*	I, NJ
C. fimbriatum	BH, I, NJ	*H. silvestris*	NM
C. hungaricum	BH, I	*H. similis*	In, I, NM
C. menzeli	BH, I	*H. vidua*	I, SHC
C. minor	NM	*Hoplolaimus galeatus*	In, NJ, NM, SHC
C. octangulare	BH, In, I, NJ	*Hoplotylus femina*	NJ
C. proclivis	BH	*Longidorus elongatus*	In, NJ, NM
C. seymouri	BH	*L. longicaudatus*	In
Criconemoides annulatum	NM	*Meloidogyne hapla*	NJ
C. axeste	BH, I	*Paratylenchus microdorus*	I, NM
C. beljaevae	SHC	*P. elachistus*	–
C. dividus	NM	*P. nanus*	NJ
C. humilis	NM	*P. projectus*	In
C. incrassatus	BH, I	*P. tenuicaudatus*	–
C. informe	NM	*Pratylenchus brachyurus*	SHC
C. lamellatus	I, NM	*P. crenatus*	NJ
C. lobatum	NJ	*P. penetrans*	NJ
C. longula	BH	*P. scribneri*	In
C. macrodorum	In, I, NJ, NM	*Rotylenchus pumulis*	SHC

C. permistus	I
C. petasus	BH, I
C. rusticum	BH, I
C. sphagni	BH, SHC
C. xenoplax	BH, I, NM, NJ
Gracilacus aciculus	NJ
G. aculentus	NM
G. audriellus	I, In, NJ
G. elegans	NJ
Helicotylenchus canadensis	NM
H. digonicus	SHC
H. dihystera	SHC, NM
H. erythrinae	SHC
H. platyurus	BH, In, I, NJ
H. pseudorobustus	In, I
R. uniformis	NJ
Trichodorus aequalis	NJ
T. christiei	NJ
Tylenchorhynchus brevicaudatus	I, SHC
T. claytoni	NJ
T. cylindricus	NM
T. dubius	NJ
T. maximus	NJ
T. silvaticus	In
Xiphinema americanum	BH, In, I, NJ, NM, SHC
X. bakeri	In
X. chambersi	In, I, NJ, NM, SHC
X. diversicaudatum	NM
X. index	NM

BH = boreal, hemlock-hardwood forest, Maine, New York, New Hampshire, Vermont (Hoffmann and Norton, 1976; Norton and Hoffmann, 1974).

In = deciduous forest, Indiana (Johnson et al., 1972).

I = oak-hickory forest, Iowa (Hoffmann and Norton, 1973; Norton and Hoffmann, 1974).

NJ = New Jersey (Springer, 1964).

NM = New Mexico (Riffle, 1968).

SHC = southern hardwood conifer forest (Ruehle, 1968b).

at one location in Nevada, but these constituted only a small part of the total nematode biomass. Siddiqui et al. (1973) listed 14 species occurring in the desert of California. Unless one knows the vegetation listed, however, interpretation is difficult. For example, *Hirschmanniella spinicaudata* is listed as being associated with *Juncus* and *Typha*, probably reflecting an oasis in the desert. While moist habitats are a part of the desert community, they should be clearly identified so that erroneous interpretations will not be made by nematologists who are not proficient in botany. Siddiqui et al. made no claim that their listings had ecological significance, however. A comprehensive taxonomic and ecological study of the desert nematode fauna is needed.

The Tundra

The tundra of North America lies north of the boreal forest except for out-liers on the mountain tops to the south. The vegetation generally consists of early flowering, rapidly growing perennials that can withstand freezing temperatures at all stages of development. Plants are usually evergreen and matted, and even trees such as alder, birch, and willow seldom attain a height of more than a few centimeters. In many parts of the tundra, plant development is affected not so much by the depth to which the soil freezes as by the depth to which it thaws. Some permafrost is called fossil ice because it has existed since the retreat of the glaciers. Grass, heaths, moss, lichens, and dwarf trees are common but vary in abundance with the environment. Regions of the continental tundra generally have high winds, little precipitation, extreme and prolonged cold, and a short growing season. The alpine tundra, especially in the United States, has more precipitation, and naturally a different day length, light quality, and light intensity. It would seem that such conditions would be inhospitable to most animal life other than Eskimos, polar bears, explorers, mosquitoes, and black flies. Nematodes of a few species survive in suitable niches, however. Unfortunately, we know little of the nematode fauna, especially the Tylenchidae, in the tundra. The explorer certainly can add to the few species that have been found.

Several studies have been made of Arctic vegetation by earlier workers such as Allgen and Ditlevsen, who recorded many dorylaimids and nonstylet nematodes but few tylenchids. Two recent studies relative to the Tylenchidae have been especially enlightening, however. Loof (1971) reported several species occurring in Spitzbergen: *Tylenchus davainei, Aglenchus bryophilus, A. costatus, Tylenchus leptosoma, T. thornei, Basiria dolichura, Geocenamus arcticus, Merlinius joctus, M. leptus, M.*

microdorus, Tylenchorhynchus magnicauda, Helicotylenchus spitsbergensis, Pratylenchoides crenicauda, and *Criconemoides hemisphaericaudatus.* Based upon collections made by R. H. Mulvey, the following are known to occur on Ellesmere Island in the Canadian Arctic: *Tylenchus aquilonius, T. cylindricaudatus, T. hazenensis, T. neodavainei, T. stylolineatus* (Wu, 1969a), *Anguina agrostis* on *Arctagrostis latifolia* (Mulvey, 1963), *Dactylotylenchus filiformis* (Wu, 1969b), *Geocenamus arcticus, Merlinius leptus, M. paraobscurus* (Mulvey, 1969b), *Deladenus durus, Hexatylus mulveyi* (Das, 1964), *Hadrodenus megacondylus, Nothotylenchus acris, N. acutus, N. attentuatus* (Mulvey, 1969a), and *Criconemoides hemisphaericaudatus* (Wu, 1965). Thus there is a diversity of Tylenchidae extant. In addition, Allen (1955) described *Merlinius alpinus* from specimens collected at 3658 m at Trail Ridge, Colorado, and I have collected it at Loveland Pass, also in Colorado. Thorne (1929) reported that *Tylenchus davainei* and *Aphelenchus pusillus* occurred at the summit (4345 m) of Long's Peak in Colorado. It will be noted that no reference is made to *Xiphinema, Longidorus,* and *Trichodorus* in these reports. It has also been noted that *X. americanum* is absent from the tundra in the Adirondacks of New York and the Green Mountains of Vermont (Norton and Hoffmann, 1974). Although information on nematodes in the tundra might seem purely academic, it does illustrate the ubiquity of some nematodes, such as *Tylenchus davainei.* It also suggests that many species common to some habitats probably would not become established in cultivated areas even if introduced.

Nematodes in Aquatic Habitats

As might be expected, most nematodes found in aquatic habitats are non-plant-parasitic (Table 3.3), but there are exceptions. Probably *Hirschmanniella* spp. are the best known tylenchids in aquatic habitats, but others such as *Helicotylenchus hydrophilus* have only been reported as occurring in or near wet environments. Certain ponds and stream banks become dry periodically and, because of environmental fluctuations, the benefit of doubt is given to some nematodes included as aquatic in the table.

Little is known about why these nematodes occur in aquatic places, and water per se might be minor in some instances. *Hirschmanniella oryzae* occurs in the aerenchyma tissue of rice, but it does not occur in the thin lateral roots where there are no air channels (Ichinohe, 1972). Nothing is known of the role that these nematodes play in natural ecosystems, but with the increased interest in aquatic biology and environmental problems, extensive work with all nematodes in these habitats is desirable.

Table 3.3. Nematodes found in aquatic habitats

Nematode	Habitat	Reference
Aorolaimus leipogrammus	Swamp soil	Sher, 1963
Bakernema variabile	Stream banks, river beds	Raski and Golden, 1965
Basiria affinis	Slough bank	Thorne and Malek, 1968
Basiroides obliquus	Pond bank	Thorne and Malek, 1968
Belonolaimus euthychilus	Sand and swamps, or alluvial soils	Rau, 1963
B. maritimus	Beaches	Rau, 1963
Criconema decalineatum	Swampy ground	Chitwood, 1957
Criconemoides discus	Slough bank	Thorne and Malek, 1968
C. raskiensis	Slough bank	Thorne and Malek, 1968
C. vernus	Just below water line	Raski and Golden, 1965
Ditylenchus radicicola	Sand marsh, low meadow, or slough bottom	Stessel and Golden, 1961; Vanterpool, 1948
Dolichodorus heterocephalus	Often in moist places	Cobb, 1914; Esser, 1964; Christie, 1952; Hutchinson et al., 1961
D. similis	Moist soil at edge of river	Golden, 1958
Helicotylenchus anhelicus	Around *Salix*, often in wet soil	Sher, 1966
H. caroliniensis	Swamp soil	Sher, 1966
H. digitatus	Cranberry marsh	Vanterpool, 1948
H. hydrophilus	Swamp soil, slough bank, pothole	Sher, 1966; Thorne and Malek, 1968; Schmitt and Norton, 1972

Species	Habitat	Reference
Hemicycliophora aberrans	Stream bank	Thorne, 1955
Hirschmanniella belli	Moist habitats	Sher, 1968
H. caudacrena	Rice soil, Everglades, aquatic plants	Sher, 1968
H. gracilis	Marshes, river banks, and other aquatic areas	Hutchinson et al., 1961; Golden, 1957; Thorne and Malek, 1968
H. marina	Marine plants	Sher, 1968
H. mexicana	Free in water	Sher, 1968; Chitwood, 1951a
H. microtyla	Around *Vallisneria americana*	Sher, 1968
H. oryzae	Rice fields	Hollis, 1967
H. spinicaudata	Edge of river	Sher, 1968
Hoplolaimus galeatus	Prairie pothole	Schmitt and Norton, 1972
H. stephanus	Swamp soil	Sher, 1963
Merlinius joctus	Prairie pothole	Schmitt and Norton, 1972
Nothanguina phyllobia	In plants growing along ditch banks or other moist places	Thorne, 1934
Psilenchus gigas	Near a slough	Thorne and Malek, 1968
Tylenchorhynchus varians	Bank of pond	Thorne and Malek, 1968
Xiphinema chambersi	Prairie pothole	Schmitt and Norton, 1972

NEMATODES AND PLANTS

Heretofore we have related nematodes to general habitats with the tacit assumption that nematode occurrence is governed greatly by plants. It is not surprising that certain nematodes parasitize some plant hierarchies more readily than others. Although in some cases we have little evidence, let us examine some of the broader relationships that appear to exist.

Most investigations on plant-parasitic nematodes have been with nematodes associated with economic crops. This is understandable, but other areas should not be totally neglected, as much information can be gathered concerning the biology of nematodes in natural ecosystems and their possible relationship to agriculture.

Fungal Relationships

A perplexing problem of the nematologist is to evaluate accurately the role of nematodes in the rhizosphere. Of the many stylet-bearing nematodes extracted from soil, some are bona fide parasites of higher plants while others are included in the "suspected plant parasite" category. Many of the latter have been demonstrated to parasitize fungi, at least under laboratory conditions, and it is possible that fungi are their primary if not sole source of enegy in the field. A partial list of mycophagous nematodes and hosts is presented in Table 3.4. A striking fact is that these nematodes are limited to relatively few genera; certain taxonomic groups are conspicuously absent. These taxonomic relationships indicate either that certain nematodes are not parasitic on fungi or that our knowledge is extremely limited, and that someday the list may become greatly expanded to include a more diverse nematode fauna. Work with enzymes, however, indicates that certain differential enzyme systems are associated with substrate and parasitic habits (Deubert and Rohde, 1971). Until there is more evidence that these more primitive nematodes can feed on higher plants, we can only assume that they are obtaining most of their sustenance from lower forms of life, with the fungi possibly being the most important substrate.

Pillai and Taylor (1967), among others, found a host preference for different nematodes. *Pyrenochaeta terrestris, Fusarium solani,* and *Alternaria solani* were rated excellent hosts for *Neotylenchus linfordi, Aphelenchus avenae, Paraphelenchus acontioides,* and a parthenogenetic and a bisexual isolate of *Ditylenchus. Phytophthora cactorum* supported moderate populations of *Ditylenchus,* but the fungus was practically a nonhost for the other nematodes. *Pythium irregulare* supported the fewest nematodes of any species tested. On the other hand, *Mortierella pusilla,* another Phycomycete, supported large numbers of *A. avenae* and moderate numbers of most other nematodes. *Rhizoctonia solani* was variable, but it supported large populations of *A. avenae* and *P. acontioides*. Although a fungal culture may support thousands of nematodes in vitro, the hyphal filaments in the soil are much more diffuse and are frequently ephemeral; consequently, populations in culture and those in soil are not comparable. Several nematodes, including *Aphelenchoides composticola, A. cyrtus, A. oxurus, A. spinosus, A. winchesi, Deladenus obesus,* and *Ditylenchus destructor* have been found in mushroom houses (Haglund and Milne, 1973; Paesler, 1957).

At least one fungus is known to repel nematodes. Townshend (1964) reported that *Sclerotium rolfsii* repulsed *A. avenae*, but this nematode and *Bursaphelenchus fungivorus* were able to reproduce on 54 or more fungi of the 59 tested. The nematodes were attracted to most fungi whether or not they were hosts. In other work, culture filtrates of *Aspergillus niger* were toxic to *Aphelenchus avenae* and the presence of the fungus in soil resulted in a reduction of nematode numbers. It was speculated that oxalic acid, which is produced by the fungus, was toxic (Mankau, 1969).

An interesting situation exists in which some phylogenetically more advanced nematodes are found in mycorrhizal roots. Species include *Criconemoides rusticum* on *Pinus echinata* (Jackson, 1948), *Hoplolaimus galeatus* and *Meloidodera floridensis* in *Pinus elliottii* and *P. teada* (Ruehle, 1962), *Helicotylenchus dihystera* and *Tylenchorhynchus claytoni* in *P. echinata* (Barham et al., 1974), *Meloidodera* sp. in *Pseudotsuga menziesii* (Zak, 1967), *Heterodera* sp. in mycorrhizal rootlets of *Pinus monticola* in Idaho (Nickle, 1960), *Meloidogyne javanica* in *P. elliottii* (Donaldson, 1967), and *Meloidogyne* sp. in *Pinus ponderosa* (Riffle and Lucht, 1966). The mycorrhizal rootlets were hypertrophied or destroyed in some instances, but it is not known whether or not there was feeding on the fungal cells. These nematodes have never been demonstrated to feed on fungi in vitro. It is known, however, that nematodes can penetrate the fungus mantle (Barham et al., 1974; Ruehle and Marx, 1971).

As might be expected, nematodes have varying degrees of feeding ability and destructiveness to fungi. Riffle (1969) reported that *Aphelenchoides cibolensis* reduced the growth of known or suspected mycorrhizal fungi. Of 40 fungi tested, growth of 19 was affected only slightly or not at all, but growth was reduced 50 to 78% with 8 fungi compared with the controls. Linear growth of *Hygrophorus piceae* and *Clitocybe crassa* was reduced to 22 and 25% respectively. Some fungi failed to grow when transferred to a fresh medium. Sutherland and Fortin (1968) reported that 10,000 individuals of *Aphelenchus avenae* per flask were probably necessary to cause a reduction

Table 3.4. Some nematodes known to feed on fungi

Nematode	Host	Reference
Aphelenchoides bicaudatus	Fungus from soil	A. A. F. Evans, 1970
A. bicaudatus	*Pyrenochaeta terrestris,* *Sporobolomyces* sp., *Hansenula saturnus*	Siddiqui and Taylor, 1969
A. bicaudatus	Reproduced in pycnidia of *Cytosporina acharii*	Schoeneweiss et al., 1967
A. besseyi	*Alternaria brassicae*	Adamo et al., 1976
A. besseyi	*Alternaria citri*	Christie and Crossman, 1936
A. besseyi	*Fusarium solani*	Huang et al., 1972
A. cibolensis	Fungi	Riffle, 1970a
A. cibolensis	*Clitopilus prunulus,* *Lepista personata,* *Russula emetica,* *Calvatia subcretacea*	Riffle, 1969
A. coffeae	*Agaricus bisporus*	McLeod, 1968
A. limberi	*Monilia sitophila*	Steiner, 1936
A. parietinus	*Neurospora sitophila*	Christie and Arndt, 1936
A. parietinus	A fungus	Bastian, 1865
A. saprophilus	?	McLeod, 1968; A. A. F. Evans, 1970
A. subtenuis	Several	Tikyani and Khera, 1969b
Aphelenchus avenae	*Agaricus bisporus*	McLeod, 1968
A. avenae	*Monilia sitophila*	Steiner, 1936
A. avenae	*Pythium arrhenomanes*	Rhoades and Linford, 1959

A. avenae	Pyrenochaeta terrestris	Hechler, 1963
A. avenae	Rhizoctonia solani	A. A. F. Evans, 1970
A. avenae	Several	Tikyani and Khera, 1969b; Townshend and Blackith, 1975
A. radicicolus	Several	Tikyani and Khera, 1969b
Bursaphelenchus lignicolus	Botrytis cinerea, Pestalotia sp.	Mamiya and Kiyohara, 1972
Deladanus durus	In decaying Pleurotus	Thorne, 1941
Ditylenchus destructor	64 species	Faulkner and Darling, 1961; A. A. F. Evans, 1970
D. intermedius	Various	Linford, 1937
D. myceliophagus	Agaricus bisporus	A. A. F. Evans, 1970; Evans and Fisher, 1969
D. myceliophagus	Agaricus campestris	Cairns, 1953
D. myceliophagus	Several imperfects and phycomycetes	Khera et al., 1968
D. triformis	Various	Hirschmann, 1962; Hirschmann and Sasser, 1955; Myers, 1965; Myers and Krusberg, 1965
Dorylaimus etterbergensis	Cephalothecium conidia	Hollis, 1957
Hexatylus fungorum	Decaying fungus (Germany)	Thorne, 1941
Neotylenchus intermedius	Alternaria citri	Christie, 1938
N. linfordi	Pyrenochaeta terrestris	Klink et al., 1970
Paraphelenchus myceliophthorus	Agaricus hortensis	Thorne, 1961
Thornia sp.	Alternaria, Pyrenochaeta	Boosalis and Mankau, 1965

in the number of mycorrhizae in *Pinus resinosa*. A minimum initial population of 4640 *Aphelenchoides cibolensis* suppressed formation of *Suillus granulatus* ectomycorrhizae with *Pinus ponderosa* (Riffle, 1975).

It has been shown that a nematode can control a plant disease. In pot tests, *Pythium arrhenomanes* produced a severe root rot, but only a mild rot developed when 125,000 *A. avenae* were added per pot. Root rot did not occur when only nematodes were added. The nematodes invaded lesions caused by *Pythium* in corn roots, but they did not invade healthy corn roots that did not contain *Pythium* (Rhoades and Linford, 1959). Although this work demonstrates that *A. avenae* can control a fungus root rot, the quantities of the nematode generally found in the fields do not approach those used in the experiment. Since *A. avenae* can feed on fungi and there is little demonstration that it increases extensively on higher plants, it probably would not reach a high enough population density to control *Pythium* root rot in the field until the root rot was greatly advanced or until other decaying debris was present.

Most nematode–fungus relationships that have been studied are relative to disease increases in combination with both organisms compared with either alone. Many of these relationships are listed and discussed in Chapter 10.

Nematodes and Algae

Relatively little work exists on the interrelationships of nematodes and algae, but green pigments are often found in the intestines of stylet-bearing nematodes, especially the Dorylaimidae, which indicates that these animals probably feed on algae. It is not known in most instances whether algae are the primary food source or are merely contaminants during feeding on other substrates. That algae can serve as a substrate, however, was demonstrated by Hollis (1957), who found that *Dorylaimus ettersbergensis* can feed on the green algae *Chlorella vulgaris* and *Tetraedron* sp. and the blue green alga *Chroococcus*, the latter being superior to the green algae in maintaining large populations of the nematode. Siddiqui and Taylor (1969) found that *Aphelenchoides bicaudatus* fed on the green alga *Stichococcus bacillaris* by pressing its lips against a cell of the algal filament and penetrating the cell with its stylet. Hollis found similar feeding habits on the globular unicellular algae *C. vulgaris* and *Chroococus*. Since *Chorella, Chroococcus,* and *Stichococcus* occur in soil, it is conceivable that these and other soil algae serve as substrates for nematodes.

Flagellates and other algae can pass through the gut of saprozoic nematodes and still be viable. Flagellates occurred internally in *Trilobus gracilis* (Bütschli, 1878) and *Diplogaster* (Ziegler, 1895), and Goodey and

Triffitt (1927) found them in the intestine of *Pristionchus lheritieri* (*Diplogaster longicauda*) inhabiting narcissus bulbs. The algae had no apparent effect on the nematode, but they seem to die shortly after being expelled from the anus. *Pristionchus lheritieri* was observed to ingest four species of algae including *Chlamydomonas reinhardi, Chlamydomonas* sp., *Ankistrodesmus* sp., and *Scenedesmus* sp. (Leake and Jensen, 1970). Many *Ankistrodesmus* were viable after passing through the digestive tract of the nematode, but few *Chlamydomonas* remained viable.

Nematodes and Lichens

Information is also meager concerning nematodes and lichens, but there are a few reports. Bastian (1865) found *Aphelenchoides parietinus* in patches of the yellow lichen *Xanthoria parietina,* and Franklin (1955) found it on the green foliose lichen *Cladonia fimbriata* var. *simplex.* Speculating that the nematode was feeding on the fungus element, Franklin attempted to culture the nematode on a fungus; she had some success in that eggs, females, and juveniles were found occassionally. Since lichens undergo periodic drying, she likewise concluded that *A. parietinus* must withstand desiccation. Allen (1955) reported that *Merlinius leptus* was found associated with moss and lichens at 3658 m in Colorado. Further work with nematodes and lichens could be of great academic interest.

Nematodes and Mosses

One might expect that mosses also could provide a niche for nematodes, and this has been amply demonstrated. Such relationships have been found from Spitzbergen (Loof, 1971) and the Canadian Arctic (Sanwal, 1965) to the warmer climates of Puerto Rico (Roman, 1965), the Canary Islands (Gadea, 1965), and Angola (Andrássy, 1963). Little has been accomplished beyond associations, however, and thus one must be cautious in interpreting reports. Although it is probable that many nematodes reported in association with mosses actually feed on the plant tissue, the possible presence of fungi that may be a source of food must also be considered. Certain nematodes are common around mosses, but others are notably absent. Whether this is due to nutritional factors or to the type of habitat, or both, is not known. Many mosses undergo periodic drying, thus probably eliminating nematodes with an inability to withstand temporary desiccation.

Some of the early collecting around mosses was by Bastian (1865) and Micoletzky (1922). In these and other early reports, members of *Monhystera, Dorylaimus,* and *Plectus* were common. Their results have been substantiated and expanded by later reports indicating that the non–

Table 3.5. Nematodes known to infest various plant parts other than roots

Nematode	Host	Reference
	Underground parts other than roots	
Criconemoides ornatum	Peanut pods	Minton and Bell, 1969
Ditylenchus destructor	Potato, tulip, iris, and many others	Faulkner and Darling, 1961; Smart and Darling, 1963
D. dipsaci	Onions, potato, among others	Thorne, 1961
Pratylenchus brachyurus	Peanut shells	Good et al., 1958
P. penetrans	Potato tubers	Dickerson et al., 1964
	Fruits, seeds, flowers, buds	
Anguina agropyronifloris	*Agropyron smithii*	Norton, 1965a
A. agrostis	Grasses	Courtney and Howell, 1952; Jensen et al., 1958
A. amsinckia	*Amsinckia intermedia*	Godfrey, 1940; Steiner and Scott, 1934
A. spermophaga	*Saccharum spontaneum*	Steiner, 1937
A. tritici	Wheat	Byars, 1920
Aphelenchoides parietinus	Cotton	Christie and Arndt, 1936
Cynipanguina danthoniae	*Danthonia californica*	Maggenti et al., 1973
Neotylenchus albulbosus	Strawberry	Steiner, 1931
Nothotylenchus acris	Strawberry	Nishizawa and Iwatomi, 1955

Stems or leaves

Anguina balsamophila	*Balsamorrhiza macrophylla*,	Thorne, 1926b
	Balsamorrhiza sagittata,	
	Wyethia amplexicaulis	
A. calamagrostis	*Calamagrostis canadensis*	Wu, 1967
A. graminophila	*Agrostis tenuis*	Goodey, 1933
A. graminophila	*Calamagrostis canadensis*	Goto and Gibler, 1951
A. pustulicola	Grass	Thorne, 1934
A. tritici	Wheat	Byars, 1920
Aphelenchoides parietinus	Cotton	Christie and Arndt, 1936
Bursaphelenchus lignicolus	Resin canals of *Pinus densiflora*	Mamiya and Kiyohara, 1972
Cynipanguina danthoniae	*Danthonia californica*	Maggenti et al., 1973
Ditylenchus dipsaci	Alfalfa, red clover, and others	Thorne, 1961
Dorylaimus intervallis	Alfalfa	Thorne and Swanger, 1936
Neotylenchus abulbosus	Strawberries	Steiner, 1931
Nothanguina phyllobia	*Solanum eleagnifolium*	Thorne, 1934
Nothotylenchus obesus	Alfalfa	Thorne, 1934
N. affinis	Alfalfa crowns	Thorne, 1941
Paraphelenchus pseudoparietinus	*Xanthium americanum*	Steiner and Buhrer, 1933
Thada striata	Alfalfa crowns	Thorne, 1941
Tylenchus polyhypnus	Rye	Steiner and Albin, 1946

Table 3.6. Feeding sites of plant-parasitic nematodes, roots unless specified otherwise

Nematode	Epidermis	Reference
Aglenchus agricola	Cabbage, cauliflower	Khera and Zuckerman, 1963
A. bryophilus	Cabbage, dill	Khera and Zuckerman, 1963
Aphelenchoides parietinus	Cotton	Christie and Arndt, 1936
Helicotylenchus dihystera	Corn	Sledge, 1959
Merlinius brevidens	*Lolium perenne*	Bridge and Hague, 1974
M. joctus	Cranberry, blueberry	Zuckerman, 1960, 1964
M. nothus	*Lolium perenne*	Bridge and Hague, 1974
M. icarus	*Lolium perenne*	Bridge and Hague, 1974
M. macrurus	*Lolium perenne*	Bridge and Hague, 1974
Paratylenchus curvitatus	Red clover, jasmine, tobacco	Rhoades and Linford, 1961a
P. elachistus	Several	Linford et al., 1949
P. pestis	Walnut	Thorne, 1943
P. projectus	Red clover, jasmine, tobacco	Rhoades and Linford, 1961a
Telotylenchus loofi	Millet	Tikyani and Khera 1969a
Trichodorus christiei	Cranberry	Zuckerman, 1961
	Tomato, rye	Rohde and Jenkins, 1957
	Red pine	Sutherland and Adams, 1964
Tylenchorhynchus claytoni	Cranberry	Zuckerman, 1961
	Alfalfa	Krusberg, 1959
T. dubius	Clovers, grasses	Klinkenberg, 1963
	Lolium perenne	Bridge and Hague, 1974
T. lamelliferus	*Lolium perenne*	Bridge and Hague, 1974
T. maximus	*Lolium perenne*	Bridge and Hague, 1974
Tylenchus emarginatus	Spruces	Sutherland, 1967b
	Sitka spruce	Gowen, 1970

Cortex

Nematode	Host	Reference
Helicotylenchus digonicus	Blue grass, lima bean	Perry et al., 1959
H. dihystera	Rose	Davis and Jenkins, 1960
	Sycamore	Churchill and Ruehle, 1971
H. pseudorobustus	Corn, tomato	Taylor, 1961
Hemicycliophora arenaria	Tomato	McElroy and Van Gundy, 1968
	Lemon	Van Gundy and Rackham, 1961
H. parvana	Bean, corn	Ruehle and Christie, 1958
H. similis	Cranberry	Zuckerman, 1961
Hoplolaimus galeatus	Pines	Ruehle, 1962
	Cotton	Krusberg and Sasser, 1956
	Pine	Ruehle, 1962
Meloidodera floridensis	*Lolium perenne*	Bridge and Hague, 1974
Merlinius icarus	*Lolium perenne*	Bridge and Hague, 1974
M. macrurus	Peach	Mountain and Patrick, 1959
Pratylenchus penetrans	Apple	Pitcher et al., 1960
P. scribneri	Amaryllis	Christie and Birchfield, 1958
P. thornei	Wheat	Baxter and Blake, 1968
Radopholus similis	Citrus	DuCharme, 1959
Rotylenchus buxophilus	Boxwood	Golden, 1956
Telotylenchus loofi	Millet	Tikyani and Khera 1969a
Tylenchulus semipenetrans	Citrus	Cohn, 1965

Root tip

Nematode	Host	Reference
Dolichodorus heterocephalus	Celery, corn, bean	Perry, 1953
Hemicycliophora arenaria	Tomato	McElroy and Van Gundy, 1968
	Cranberry	Zuckerman, 1961
	Blueberry	Zuckerman, 1964
H. similis	Lettuce, marigold, cabbage	Klinkenberg, 1963

55

Table 3.6. *(Continued)*

Nematode		Reference
Longidorus africanus	Grape, nettle, marigold	Cohn, 1970
Merlinius joctus	Cranberry	Zuckerman, 1960
	Blueberry	Zuckerman, 1964
Telotylenchus loofi	Millet	Tikyani and Khera, 1969a
Trichodorus christiei	Tomato	Rohde and Jenkins, 1957
	Cranberry	Zuckerman, 1961
T. proximus	St. Augustine grass	Rhoades, 1965
T. viruliferus	Apple	Pitcher and McNamara, 1970
Tylenchorhynchus martini	Rice	Hollis and Fielding, 1958
T. maximus	*Lolium perenne*	Bridge and Hague, 1974
Xiphinema bakeri	Douglas fir, Sitka spruce	Sutherland, 1969
X. index	Grape, fig, rose (?)	Radewald and Raski, 1962b
	Herbaceous plants	Cotten, 1973
	Stele	
Heterodera trifolii	Clover	Mankau and Linford, 1960
Hoplolaimus galeatus	Cotton	Krusberg and Sasser, 1956
Meloidogyne naasi	Wheat	Siddiqui and Taylor, 1970a
Rotylenchulus reniformis	Cotton	Birchfield, 1962
Tylenchorhynchus claytoni	*Pinus echinata*	Barham et al., 1974
	Variable	
Criconemoides curvatum	Carnation	Streu et al., 1961
C. lobatum	Zoysia grass	A. W. Johnson and Powell, 1968
C. xenoplax	Peach	Thomas, 1959
Xiphinema diversicaudatum	Petunia hybrida	Trudgill, 1976

plant-parasitic nematodes seem to predominate (Brzeski, 1962a, 1962b, 1963a; Ditlevsen, 1927; Eliava, 1966; Gadea, 1964; Overgaard-Nielsen, 1948; Steiner, 1916). Many nematodes associated with mosses are regarded as bacterial feeders and predators, and most of the tylenchids and aphelenchids associated with mosses are of the primitive type. Among them are *Aglenchus machadoi, Tylenchus agricola, T. davainei, T. filiformis, T. bryophilus, T. leptosoma, T. ditissimus, Aphelenchoides parietinus, A. villosus, A. arcticus, A. appendurus, A. helophilus, A. montanus,* and *A. parasaprophilus* (Andrássy, 1963; Bastian, 1865; Brzeski, 1962a, 1962b, 1963b; Ditlevsen, 1927; Eliava, 1966; Gadea, 1964; Loof, 1971; Sanwal, 1965; Singh, 1967; Steiner, 1916; Zullini, 1970, 1971). Less primitive Tylenchoidea have been reported fewer times. Allen (1955) reported *Merlinius leptus* to be associated with moss and lichens at 3658 m in Colorado, and Loof (1971) mentioned *Geocenamus arcticus* as occurring around mosses in Spitzbergen. Other reports include: *Helicotylenchus thornei* from moss on tree trunks in Puerto Rico (Roman, 1965); *Rotylenchus robustus, Criconema menzeli, Criconemoides annulifer,* and *C. sphagni* as occasional occurences in peat moss in Poland (Brzeski, 1962a, 1962b); *C. sphagni* on moss in the United States (Hoffmann, 1974); and *Hemicycliophora typica* in Italy (Zullini, 1971).

Although most studies with the Tylenchida and mosses have led to reports of associations, there is increased interest in exploring the relationships of nematodes and mosses. Zullini (1970) examined mosses at ten different stations ranging from 1340 to 2150 m in the central Alps. With the Baermann funnel method of extraction, only 2% of the nematodes recovered were found to be plant feeders. *Aphelenchoides parietinus* was classified as a more xeric species than *Tylenchus davainei* on the basis of the moss wetness. Survival as well as viability might be a factor in habitation.

TISSUE SELECTION IN HIGHER PLANTS

Once a nematode has become compatible with the general climatic regime, the controlling environment, including the host's morphology, becomes more microclimatic. Most plant nematodes parasitize the roots, but many feed on other plant parts both above and below ground. A few, such as *Ditylenchus dipsaci* on teasel, are known to destroy the entire plant. Listed in Table 3.5 are many nematodes that parasitize or infest plant parts other than roots. The list is not meant to be all-inclusive. Some of the better known associations, such as many *Aphelenchoides* species, are omitted. Some associations have been reported infrequently and may be only incidental. This might indicate a lack of observation and research, especially in instances involving noneconomic plants. A scan of this list quickly makes it evident, however, that certain groups of nematodes are more prevalent in certain plant parts and also that the bulk of the well-known root parasites are absent. In nearly all instances, the primitive parasites prevail in the above-ground parts, whereas the below-ground nonroot parts are parasitized by several nematodes representing a cross section of parasitic types. Although such information is useful in predicting what you might expect to find, one should always look for the unexpected. There are instances where root-knot and cyst nematodes have been reported feeding above ground, but these are usually in the greenhouse under humid conditions or some unusual circumstance. Also *Anguina agropyronifloris* (Norton and Sass, 1966) and *A. tritici* (Byars, 1920) are capable of feeding on root tissue, but apparently this is of no consequence in the life cycle of the nematode.

Segregation also occurs among the root feeders. Most epidermal feeders of the Secernentea generally have small stylets, a well-developed basal bulb, and median bulbs that are either degenerate or are not oversized (Table 3.6). The Criconematidae, which are the most advanced of the migratory Secernentea, often have large stylets, an enlarged valvular apparatus, and a reduced terminal bulb. McElroy and Van Gundy (1968) correlated feeding processes, nematode morphology, and developmental trends in parasitic specialization. Most feeding observations have been made under highly artificial conditions in the laboratory and thus some variance is probable under more natural conditions. All in all, however, laboratory observations are at least not contradicted by field phenomena. Evidently, feeding at a site is not a random phenomenon but is governed by a multitude of interacting factors involving both host and parasite. There is evidence that, at least in one instance, the feeding site is related to breeding. Pitcher and McNamara (1970) reported that the proportion of *Trichodorus viruliferus* that developed eggs was higher (80%) at the root tips of mature apple trees than in samples of this nematode taken at random (18%). They concluded that breeding follows aggregation at the root tip rather than the reverse.

COMMUNITIES

Every biotic community is characterized by periodic
fluctuations in the relative abundance of its constituent
species

A community is an assemblage of individuals; it is a container of information. It may be large or small and may be composed of individuals of identical or different species. These may be plants or animals, autotrops or heterotrops. They live, they die, they utilize what went before them. Many provide sustenance for those that follow. The organisms may be producers, consumers, predators, or decomposers. No matter what their nature, they all have one thing in common—they operate within the framework of their environment, and thus they help structure the community in which they live. With only normal climatic changes, cyclic patterns are set. If sufficient information is available, one can predict the end result within the realm of tolerance. Because long-term changes in undisturbed habitats come slowly, an observation some years hence would be much the same as today. Part of the cyclic pattern is due to seasonal climatic changes and part to the results of parasitism or other forms of competition. The population of nematodes that feed on roots may increase in size and density. Plant tissue necrosis initiated by nematodes may be abetted by fungi, bacteria, and other organisms. Rootlets die, and the nematode population declines. Carcasses of dead nematodes provide sustenance for other microorganisms and, indirectly, even nutrition for the host on which they were feeding. New rootlets form, and the surviving nematodes are quick to parasitize them. The cycle begins anew. The steady state, as man sees it, has been maintained. The system of negative feedbacks has resulted in relative stability and, thus, persistence of species.

DIVERSITY, STRUCTURE, AND STABILITY ⸻⸻⸻⸻⸻

Stability implies resistance to change and thus continuance of the species or community. The stability of a nematode community is governed by the stability of the abiotic environment and the interactions among the biotic components, including the host. Much work exists on nematode associations with other organisms, especially fungi, on a one-to-one basis, or interacting with three or four organisms (see Chapters 9 and 10). Most of these mixed associations have been studied relative to disease incidence and severity. We are just beginning to document the community as a whole, and information is sparse. We know that other organisms in the soil are important because of the ease with which some nemtodes increase in the greenhouse, where competition is absent. Yet the same organism might not increase greatly in nonsterilized soil. Most ecology textbooks relate diversity, structure, and stability very closely, although most authors admit that a cause-and-effect relationship is difficult to prove. Few nematode communities have been analyzed in this manner, but from observations it seems that the associations are not fortuitous. The holistic approach in community analysis is desirable but is not often practical. Thus, most studies are made at defined trophic levels.

Diversity

Diversity sometimes is defined as the number of species of a given taxon or the number of taxa in a community. According to this concept, a community with nine species of nematodes is more diverse than a community with only five species. Most students of diversity include quantitative values such as numbers or biomass as a diversity measure along with qualitative values. If two species dominate a community that consists of nine different species, the community is less diverse than if all nine species were equally proportioned. The Shannon-Weiner index is one of the most commonly used diversity measures, but its use alone may be an over-simplification of the real situation. This index, however, is a good place for the student to obtain an introduction to the subject. The reader is referred to Wilson and Bossert (1971) and Pielou (1975) for elementary and advanced treatments of diversity.

Diversity studies have not been used much in nematology except for the general observations that more species usually occur in some areas than in others. Yeates (1970) suggested that in listing nematodes the total number of species should be given in descending order of abundance. The author also should indicate the least number of species required to include 75% of the individuals, and those species that individually make up 2% or more of the fauna. It would be worthwhile to pursue diversity further, especially since diversity, structure, and stability seem interrelated.

Structure

The importance of diversity can be seen readily as we discuss structure. Structure is thought of in different ways. Faunistic composition sometimes is referred to as faunistic structure. Different species may occur at different vertical levels; this is called species stratification or vertical structure. As examples, Richter (1969) found more *Trichodorus viruliferus* at the 10 to 40 cm depth range, but *T. pachydermus* occurred mainly between 30 and 60 cm deep. Yuen (1966) found that *Helicotylenchus vulgaris* and *Rotylenchus pumilus* had different stratification patterns in the same habitat. There is also vertical structure in which one species has a numerical or biomass stratification. This probably occurs in most nematode populations and is illustrated graphically by H. Ferris and McKenry (1974) for *Meloidogyne* spp. and *Xiphinema americanum* (Figures 1.3 and 1.4). Trophic groups can be used as a life form structure. Community structure sometimes is used for faunistic composition. But, ecologically, structure most often refers to the pattern of interrelationships of organisms.

Mixed populations of nematode species can constitute a community. With no futher thought than this, one might consider that a community could represent a static or even fortuitous grouping. But we usually assume that natural populations have some interrelationship; the very nature of speciation and evolution would seem to dictate it. Parasitism dictates an interaction between at least two members of the community, namely the host and parasite. Most ecologists would agree that without an interrelationship the community is not structured. A community may be loosely structured or strongly structured depending on the degree of interaction among its members. Groups of species may be structured differently in time or location because of environmental differences and age of the nematodes. A developing population will have a different structure than a declining one. Species need not coexist or be in close proximity to be interacting; they may operate at a distance and over time. For example, morphological or physiological changes that occur in galls caused by the root-knot nematode may prevent other nematodes from feeding later on that tissue. The gall also may be a more favorable substrate than nongalled tissue for some fungi (Powell and Nusbaum, 1960).

As far as we know, most plant-parasitic nematodes are not dependent on other organisms except for the host, but other organisms may be partly or totally dependent on them. Nematodes modify their environment, and changes in community structure are bound to follow. Interactions start as soon as a nematode enters a community. The nematode either finds a niche in that community or it will not become established and eventually will be excluded. Because of the complex nature of the soil environment, probably no two communities are exactly alike. The plant-parasitic nematodes are of the greatest interest to us, but because of interrelationships with other components of the ecosystem the holistic approach should be of concern, at least within a given trophic level.

Stability

Diversity and structure generally affect stability. The greater the number of species and the greater the number of individuals within those species, the more interactions among them are possible. It follows that the more interactions there are, the less likely it is that one species will tend to dominate. Also, the more interactions there are, the less likely it will be that the removal of one species will greatly alter the community structure. The fewer the species in a community, the more loosely structured that community probably will be, and thus the less stable. A plant-parasitic nematode community may reach a sizable population in annual crops by the season's end, depending on the residual density from a preceding crop and the environmental conditions during the ensuing year. Different host genotypes, various types of plant spacings and soil tillages, along with various periods during which the soil is fallow or planted with a nonhost keep a nematode community in constant change. It is doubtful if a community can really become stabilized enough under these artificial conditions to be considered mature.

The greatest disturber of the steady state is man himself. With the clearing of the forests and the breaking of the sod, new environments were created and many nematodes could not survive. Some species that were living a marginal existence in undisturbed areas were favored by the new environment, and they flourished. But has a new steady state arisen? If so, humans have made it highly tenuous because a natural equilibrium is not attained. As long as people set the boundaries, a kind of steady state may exist in that the same species may occur year after year. But once the cycle is broken by changing management practices, progression toward a new steady state begins. Whether or not a new steady state is attained depends on the frequency and effectiveness of changes in management practices. Frequent alterations of the environment are basic to ecological nematode control. There are a few instances in plant pathology in which a disease has not become worse or has even declined by monocropping. But this seems to be the exception and not the rule.

A greater stability is apt to result in noncultivated agroecosystems, such as pastures, hay fields, and unplowed orchards, because of less disturbance by man. The greatest stability is generally in natural areas, which usually have a greater diversity of species than cultivated areas and are subject to relatively few disturbances (S. R. Johnson et al., 1972; Schmitt and Norton, 1972; Yuen, 1966).

The forces acting on stability determines the size of the population. Internal forces of a system are those that are restricted within the existing parameters of soil, such as pH, organic matter, and other physical and chemical factors. External forces such as those imposed by man in agroecosystems are apt to produce more drastic and less predictable changes than those in natural areas. Monoculture, cultivation, herbicides, irrigation, and other farming practices narrow the ecosystem and reduce its stability by removing some of the restraints compared with noncultivated land. Stability studies are complex, and few good quantitative data over long periods are available for plant nematodes.

Long-term stability and changes with parasitic nematodes doubtless are greatly influenced by the stability of the host. I know of no data in nematology where this factor has been followed for many years. We know that there is a persistence of species by habitat, both natural and in agroecosystems. Inherent in stability or nonstability is the degree of fluctuation of the organisms, which varies greatly with species and habitat. Different degrees of fluctuation of *Helicotylenchus pseudorobustus* and *Xiphinema americanum* occurred under different tillage systems in a single corn field (Thomas, 1978) (Figure 4.1).

TROPHIC GROUPS

Nematodes are often classified according to the substrate upon which they feed. Various groupings have been used (Banage, 1963; Lee, 1965; Overgaard-Nielsen, 1949; Paramonov, 1962; Wasilewska, 1971; Winslow, 1960), but most include the following in some form.

Microbivorous and *saprophagous*—feed mainly on bacteria and products of decay. (There is no such thing as a saprophytic nematode in spite of the term's frequent occurrence in the literature.)

Plant-parasitic—strictly speaking, feed on any plant, including fungi, algae, and other primitive forms. Often implied in the connotation is reference only to bryophytes, pteridophytes, and spermatophytes. Since so many nematodes are known to feed on fungi, it is best to keep the plant hierarchies separate.

Mycophagous—use fungi as a substrate.

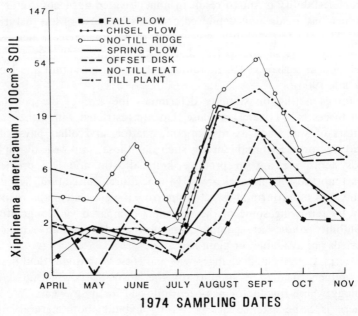

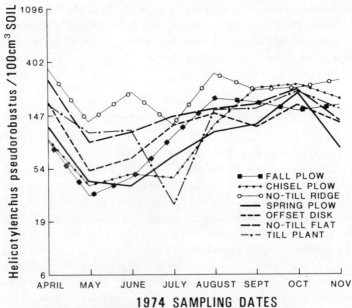

Figure 4.1. Populations of *Xiphinema americanum* (top) and *Helicotylenchus pseudorobustus* (bottom) under seven different tillage systems in one corn field. (From Thomas, 1978.)

Carnivorous (called *predators* by some authors)—feed on animals such as protozoa, other nematodes, rotifers, tardigrades, and other animals.

Omnivorous—feed on a diversity of substrates.

Food unknown—If we were realistic, many more species would be included here than is frequently acknowledged.

Each grouping can be subdivided to suit the needs of the investigator. As Wasilewska (1971) pointed out, there are no clear-cut boundaries between trophic groups in many instances, and it is difficult to decide in which ecological grouping some species belong. Placement usually is based on morphological features and the known feeding habits of related species. This is logical and reasonable to a point, but uncertainties exist. In the first place, especially among the more primitive tylenchids, the feeding habits are not always known and the possession of a stylet does not always indicate feeding on a higher plant. Some species of the Tylenchinae feed on roots of spermatophytes (Table 3.6), and some feed on fungi (Table 3.4). The same is true for members of *Aphelenchoides, Ditylenchus,* and other genera. If a nematode might fit into more than one category, it often is placed in the apparently predominant group based on the direct or indirect evidence. A simple example of trophic levels relative to plant-parasitic nematodes is given in Figure 4.2.

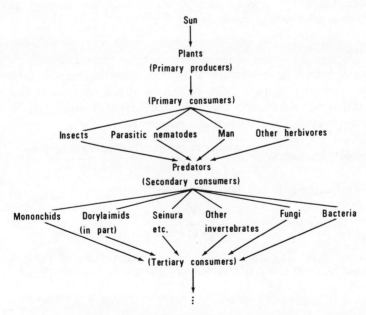

Figure 4.2. Trophic levels relative to nematodes.

SOME SIMPLE METHODS OF COMMUNITY ANALYSIS _____

It is difficult to assimilate all the information necessary for analysis, and pragmatically we select for study those facets that might be important for nematodes. Often the facets will be based on previous knowledge, theoretical consideration, ease of instrumentation, educated guesses, or scientific curiosity. Whatever the reason for undertaking an investigation, samples are processed, examined for nematodes, and recorded. The investigator must then summarize the data so they can be meaningful to himself and to others. The usefulness of this summary depends upon the thoroughness of investigation and the manner of presenting the data. Reporting can be accomplished in several ways. Each method usually has its merits. Here we shall discuss only some of the simpler methods of presenting data in ecological work. Mathematical and statistical ecology are becoming more important in biological studies, but advanced techniques, and even some simple ones, have been used little in nematode ecology. Some are still in the experimental stage, and there are advocates pro and con to various uses.

The simplest method of presenting data is to list the nematodes found in an area or crop, frequently within specified political boundaries. Such a list provides no information concerning frequency, abundance, or habitat relationships, except sometimes of associated plants. Thus a mere listing of species is of limited long-term ecological merit. Surveys are not to be discredited, but annotations on habitats, soils, frequency, density, and other parameters add much more meaning.

Frequency

Frequency, that is, how often a species occurs among samples, is an improvement over a mere listing of species. It is a measure of distribution uniformity, not abundance; one individual counts as much as a thousand. Absolute frequency is expressed as a percentage:

$$\text{absolute frequency} = \frac{\text{number of samples containing a species}}{\text{number of samples collected}} \times 100$$

Relative frequency is calculated as follows:

$$\text{relative frequency} = \frac{\text{frequency of species}}{\text{sum of frequency of all species}} \times 100$$

Assume that the data for plant-parasitic nematodes in Table 4.1 were collected from 93 samples. The calculation for absolute frequency of *Helicotylenchus pseudorobustus* is

$$\frac{58}{93} \times 100 = 62.4\%$$

The relative frequency for the same species equals

$$\frac{62.4}{273.1} \times 100 = 22.8\%$$

Or it also could be calculated as $(58/254) \times 100$. Relative frequency puts the total frequency on the basis of 100%, but the frequency of occurrence of one nematode relative to another does not change whether absolute or relative frequency is used. *Xiphinema americanum* is not quite seven times as frequent as *T. nudus* in either instance. Although the above examples incorporate only plant-parasitic nematodes, frequencies of species relative to all nematodes or groups of nematodes could, of course, be made.

Density

Density (abundance) is a quantitative measure of entities in a sample or a mean for a group of samples per unit of soil—for example, 30 *Hoplolaimus galeatus* per 100 cc of soil is a measure of absolute density. We may also wish to know how many individuals of a species are in a sample relative to the total mean number of individuals for all species. This is relative density and is usually expressed as a percentage.

$$\text{relative density} = \frac{\text{number of individuals of a species in a sample}}{\text{total of all individuals in a sample}} \times 100$$

If we assume that the mean densities of the nematodes in all 93 samples as used in Table 4.1 are in the middle column of Table 4.2, then the relative densities are those in the right-hand column. The calculations for *H. pseudorobustus* would be $(131/247) \times 100 = 53.0\%$.

Table 4.1. Frequencies of nematodes occurring in 93 corn samples

Nematode	Number of samples containing species	Absolute frequency (%)	Relative frequency (%)
Pratylenchus hexincisus	87	93.5	34.2
Xiphinema americanum	62	66.7	24.4
Helicotylenchus pseudorobustus	58	62.4	22.8
Quinisulcius acutus	21	22.5	8.2
Hoplolaimus galeatus	17	18.3	6.7
Tylenchorhynchus nudus	9	9.7	3.6
Total	254	273.1	99.9

Table 4.2. Mean absolute and relative densities of nematodes per unit of soil in 93 samples

Nematode	Absolute density	Relative density (%)
Helicotylenchus pseudorobustus	131	53.0
Xiphinema americanum	38	15.4
Quinisulcius acutus	27	10.9
Pratylenchus hexincisus	19	7.7
Hoplolaimus galeatus	18	7.3
Tylenchorhynchus nudus	14	5.7
Total	247	100.0

Prominence Value

Since density and frequency provide different information about populations, authors in some disciplines have combined the two in hopes of having a figure that would relate aspects of each. One such arrangement is the prominence value (PV) of Beals (1960):

$$PV = \text{density } \sqrt{\text{frequency}}$$

The rationale was that density was more important than frequency and that frequency modified density through space. Although Beals used it for studying bird populations, the emphasis on density is of practical value to nematologists, since, if we can reduce nematode densities below injury levels, frequency often is of little importance. If we continue with our examples of nematodes in Table 4.1 and 4.2, the PV values are as listed in Table 4.3. Note that *P. hexincisus* and *Q. acutus* have exchanged places in rank from those of densities due to modifications by the frequencies.

Table 4.3. Prominence values for plant-parasitic nematodes based on absolute density and absolute frequency

Nematode	Frequency	Density	PV
Helicotylenchus pseudorobustus	0.62	131	103.1
Xiphinema americanum	0.67	38	31.1
Pratylenchus hexicisus	0.94	19	18.4
Quinisulcius acutus	0.23	27	12.9
Hoplolaimus galeatus	0.18	18	7.6
Tylenchorhynchus nudus	0.10	14	4.4

Although we may be getting closer to determining the importance of nematodes in a sample or group of samples than by just examining frequency or density alone, it may be that the most prominent nematodes are not the most important. Among other parameters, the biomass differs greatly among species. This is used in the importance value which, along with biomass, is discussed in the next section.

Biomass

During the host–parasite interaction, there is a certain energy flow from the plant to the nematode. Although numbers of nematodes have been used extensively in reporting results, biomass relates to the energetics of the ecosystem more than numbers do. Different methods of figuring biomass exist, but the equation of Andrássy (1956) is commonly used:

$$G = \frac{a^2 b}{16 \times 100,000}$$

where G equals biomass in micrograms, a is the greatest body width, b equals the body length, and 16 is a previously determined empirical value. A discussion of the derivation and comparisons among nematodes is given by Andrássy (1956). Biomass figures often are based on the adult female even though all individuals of a species are included. This results in an over-weighted value, because in an expanding population the juveniles greatly outnumber the adults. The ideal situation would be to obtain an average biomass figure for every juvenile stage plus the adult stage. This is sometimes difficult and certainly laborious. An average for all stages present is preferable to only adult figures, however. Yeates (1973) found that the geometric growth rate for biomass varied within the species (Figure 4.3) and was generally logarithmic or nearly so.

Adult biomass figures for a few plant-parasitic nematodes are listed in Table 4.4. Since infraspecific geographical variants occur, it is best to make measurements for each study, although differences may be minor in many instances. Biomass figures of total or groups of nematodes vary with habitat (Table 4.5). As would be expected, biomass figures for any trophic group will change seasonally, as Wasilewska (1974) found with nematodes associated with rye. Figures will also vary among authors because of differences in inclusion of different genera within the compilation. For example, Wasilweska (1971) placed *Aglenchus, Boleodorus,* and *Tylenchus* with parasites of higher plants, whereas I would place them in the mycophagous group even though a few are known to parasitize higher plants. Unfortunately, we do not know the feeding habits of many of these more primitive tylenchids.

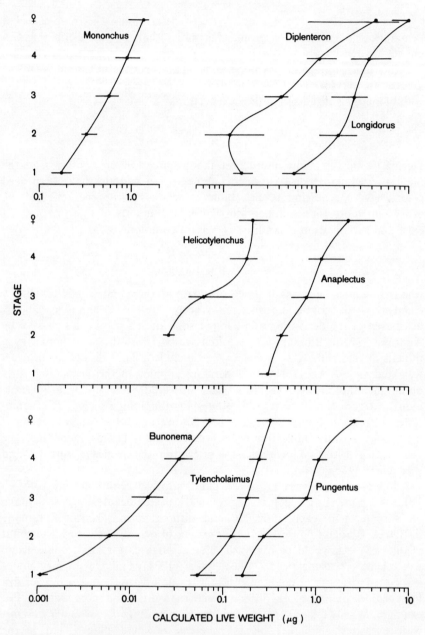

Figure 4.3. Increase in biomass of eight soil nematodes during the life history. (From Yeates, 1973, p. 720. Courtesy New Zealand Journal of Science.)

Table 4.4. Biomass of adult nematodes

Nematode	Number and sex	Biomass (μg)	Source for measurement
Aglenchus costatus	F	0.107	Thorne, 1961
Criconemoides curvatum	F	0.208–0.337	Raski, 1952
C. xenoplax	F	0.250	Thorne and Malek, 1968
Helicotylenchus digonicus	20F	0.147–0.282	Sher, 1966
H. exallus	20F	0.211–0.290	Sher, 1966
H. exallus	10M	0.150–0.194	Sher, 1966
H. hydrophilus	20F	0.317–0.507	Sher, 1966
H. hydrophilus	10M	0.279–0.309	Sher, 1966
H. leiocephalus	20F	0.314–0.412	Sher, 1966
H. platyurus	F	0.410	Thorne and Malek, 1968
H. platyurus	8F	0.510–0.420	Sher, 1966
H. platyurus	F	0.630–0.490	Perry, Darling, and Thorne, 1959
H. pseudorobustus	20F	0.180–0.300	Sher, 1966
H. pseudorobustus	F	0.190	Thorne and Malek, 1968
Hemicycliophora obesa	1F	4.247	Thorne, 1955
H. penetrans	1F	1.000	Thorne, 1955
H. penetrans	1M	0.408	Thorne, 1955
Hoplolaimus galeatus	20F	1.190–3.950	Sher, 1963
H. galeatus	10M	1.160–2.320	Sher, 1963
Longidorus breviannulatus	14F	7.460	Norton and Hoffmann, 1975
Paratylenchus projectus	83F	0.066–0.064	Jenkins, 1956
Pratylenchus alleni	10F	0.065	V. R. Ferris, 1961
P. alleni	10M	0.047	V. R. Ferris, 1961
P. hexincisus	95F	0.080–0.280	D. P. Taylor and Jenkins, 1957
P. neglectus	F	0.057–0.166	Sher and Allen, 1953
P. scribneri	37F	0.158–0.259	Loof, 1960
Psilenchus hilarulus	F	0.350–1.060	Thorne and Malek, 1968
Quinisulcius acutus	3F	0.160–0.150	Allen, 1955
Q. acutus	7F	0.140–0.200	Knobloch and Laughlin, 1973
Trichodorus christiei	20F	0.270–0.560	Allen, 1957
T. christiei	7M	0.530–0.420	Allen, 1957
Tylenchorhynchus maximus	12F	0.430–0.777	Allen, 1955
T. nudus	2F	0.25	Allen, 1955
T. silvaticus	10F	0.554–0.574	V. R. Ferris, 1963
T. silvaticus	10M	0.317–0.506	V. R. Ferris, 1963
Tylenchus davainei	F	1.000–1.121	Thorne, 1961
Xiphinema americanum	13F	1.610–0.810	Knobloch and Laughlin, 1973

Table 4.5. Some figures on biomass by habitats

Habitat	Biomass (g/m²)	Comments	Authority
Colorado desert, USA	0.09	To 20 cm depth	Freckman et al., 1975
Irrigated field soil, USA	0.16	To 20 cm depth	Freckman et al., 1975
Mojave Desert, USA	0.17	To 20 cm depth	Freckman et al., 1975
Nevada desert, USA	0.03	Between plants	Freckman et al., 1974
Nevada desert, USA	0.12	Under plant canopy	Freckman et al., 1974
Rangeland	0.40–0.50	To 40 cm depth	Smolik and Rogers, 1976
Mountain pasture	about 0.30	Annual mean	Wasilewska, 1975
Rye	0.14–0.19	Mean of 8 months	Wasilewska, 1974
Beech forest	4.05	To 25 cm depth	Volz, 1951
Oak forest	15.15	To 25 cm depth	Volz, 1951
Woodland	2.13	Many dorylaimids	Johnson et al., 1974
Woodland	1.04	Few dorylaimids, many *Helicotylenchus platyurus*	Johnson et al., 1974
Woodland	1.73	Mean of 18 sites	Johnson et al., 1974

Importance Value

Frequency, density, and biomass are among the parameters that encompass the importance of a nematode species in the community. Curtis (1959) popularized the concept of importance value when working with plant communities, and the value has been used in other disciplines. Curtis defined the importance value (IV) as relative frequency + relative density + relative dominance. Since dominance is a vague term with nematodes, relative biomass is substituted here for relative dominance. The absolute biomass, relative frequency, relative density, relative biomass, and the importance value of the species that we have used in our previous examples are provided in Table 4.6. Note that the ranks of some nematodes change from previous rankings when we consider absolute biomass, relative biomass, and importance value.

As valuable as prominence and importance values might be, other aspects of nematodes, namely parasitism, might mask or take precedence over these values when pathogenicity is involved. How do we transfer some biological properties into mathematical numbers? That is, how can we weigh the bio-

Table 4.6. Absolute biomass and importance values of selected nematodes

Nematode	Absolute biomass (n)	Biomass/ nematode μg	Total biomass/ species μg
Xiphinema americanum	38	0.995	37.8
Helicotylenchus pseudorobustus	131	0.257	33.7
Hoplolaimus galeatus	18	0.792	14.3
Quinisulcius acutus	27	0.091	2.5
Tylenchorhynchus nudus	14	0.125	1.8
Pratylenchus hexincisus	19	0.072	1.4
Total	254	2.332	91.5

	Relative frequency	Relative density	Relative biomass	Importance value
Helicotylenchus pseudorobustus	22.8	53.0	36.8	112.6
Xiphinema americanum	24.4	15.4	41.3	81.1
Pratylenchus hexincisus	34.2	7.7	1.5	43.4
Hoplolaimus galeatus	6.7	7.3	15.6	29.6
Quinisulcius acutus	8.2	10.9	2.7	21.8
Tylenchorhynchus nudus	3.6	5.7	2.0	11.3
Total	99.9	100.0	99.9	299.8

logical properties of a nematode to provide a reasonable assessment of its biological impact? Should an endoparasite be weighed more heavily than an ectoparasite? Should an endoparasite that feeds internal to the endodermis be weighed more heavily than one that feeds in the cortical parenchyma? What are the roles of toxins and host resistance in the scheme? These are problems that need study to see if weighed figures can be added to or included in the importance value. So far, no one has proposed such a scheme for nematodes. There has been much work on this aspect with individual nematodes, and some on combinations of nematodes, but little on communities as a whole.

This raises the question of whether or not the total nematode damage is any greater in mixed plantings than it is in monocropping. Do we know whether nematodes in mixed plantings create a number of little problems rather than a few big ones? I know of no data that answer this question. Experience with other ecosystems tells us that a dilution of susceptible germ plasm mitigates the spread or devastation of a disease. But what is the effect of all parasites and diseases in the ecosystem, especially those below ground? Has the system of feedbacks in multiplant ecosystems relegated parasitism by several species to a minor role in yield reduction? Probably much depends on whether or not the host–parasite interaction is strong, such as might be true with a sizable population of *Hoplolaimus galeatus,* or weak, as with a moderate or weak pathogen of an epidermal feeder. If, in mixed cultures, more weak pathogens are operating by a shift in germ plasm and there are fewer strong pathogens than occurred under monoculture, then the overall amount of pathogenicity probably would be less. The reverse could also be true. But this is theoretical and needs to be tested by experimentation. Doubtless, intensive cultivation has removed many buffers that occur in natural and mixed systems. It is the scientists' task to find answers to dilemmas that man has created and to educate people on how to keep losses to a minimum in environmentally compatible ways.

Index of Similarity

An index of similarity might be used if one wishes to compare the nematode fauna of two areas or test sites. One of the earliest in ecology, by Jaccard (1912), is based on presence-absence and is the ratio of the number of species common to both areas to the number of all species found in both areas. This has been modified for quantitative use also. It is often expressed as percent, but it does not have to be, especially when used in a 2×2 contingency matrix.

$$IS_j = \frac{\text{number of species common to two areas}}{\text{total number of species in the two areas}}$$

The formula has been written in different ways, but the results are the same. For example,

$$IS_j = \frac{c}{a + b + c}$$

where c equals the number of species common to both areas, a equals the number of species found only in area a, and b equals the number of species found only in area b.

Sorensen's (1948) method, a modification of Jaccard's equation, is more widely used. Again, it was used originally for presence-absence when comparing two areas, but it also has been modified for quantitative purposes.

$$IS_s = \frac{c}{\frac{1}{2}(a + b)} \quad \text{or} \quad IS_s = \frac{2c}{a + b}$$

where c equals the number of species that the two samples or areas have in common, and a and b are the number of species in samples or area a and b, respectively. Sorensen's index has the advantage compared with Jaccard's in that the actual common species (numerator) are related to the theoretically possible common species (denominator). The philosophy behind these indices is discussed by Sorensen. Both Jaccard's and Sorensen's indices form the basis for much of the modern ecological approaches.

If several plots, samples, or habitats are compared, a table can be constructed for easy visualizing of the plots that are most similar or dissimilar based on presence-absence. As an example, let us take another simple hypothetical situation where five nematodes are compared in five habitats. Assume the data in the upper part of Table 4.7. A 2×2 contingency table can be constructed based on a similarity index, Sorensen's in this instance. Similarity indices are first obtained for habitats A and B, followed by comparisons of habitats A and C, A and D, A and E, B and C, and so on until all comparisons have been made. These figures are inserted in the lower part of Table 4.7. If the same species occur in two compared habitats, as in C and D, the similarity index is 1.0 for presence-absence. It does not mean that all species that occur in all habitats occur in C and D, but that the species that do occur are the same. The species in habitats A and E were the most dissimilar. An index of zero would indicate no similarity. Multiplication by 100 could be used for percentages.

The above tells us that there are qualitative differences among habitats but that in presence-absence all nematodes count equally. Beals (1960) and Bray and Curtis (1957) have used a modification of Sorensen's equation:

$$C = \frac{2w}{a + b}$$

Table 4.7. Number of nematodes per 100 cc soil in different habitats

Nematode	Habitat				
	A	B	C	D	E
1	20	0	10	30	0
2	180	160	40	90	60
3	490	520	110	270	20
4	70	30	10	70	0
5	40	10	0	0	0
Total	800	720	170	460	80

Habitat	IS_j				
	Habitat				
	A	B	C	D	E
A		.889	.889	.889	.571
B			.750	.750	.667
C				1.000	.667
D					.667
E					

where w is the sum of the least quantitative values that the compared habitats have in common, and a and b are the total densities in habitats A and B, respectively. Any quantitative value such as density, PV, or biomass can be used. When the density data in Table 4.7 are applied to the foregoing equation, the similarity indices are those in Table 4.8. The calculation for comparing habitats B and D are

$$\frac{2 \times (0 + 90 + 270 + 30 + 0)}{720 + 460} = .661$$

The procedures just described are simple, and, unless too many nematodes or plots or habitats are used, calculations can be done by hand or with a simple desk calculator. Advanced programs require computer help. These advanced techniques have been used little in plant nematology and they will only be mentioned in passing. The techniques used in nematology have been borrowed from other disciplines. Books have been written on mathematical ecology, and the student should obtain formal training if possible.

Clustering

Similarity indices have been used in ordination of nematode populations (S. R. Johnson et al., 1972, 1973, 1974), cluster analysis (Schmitt and Norton,

Table 4.8. Index of similarity for density of nematodes in different habitats

Habitat	Habitat				
	A	B	C	D	E
A		.908	.351	.714	.182
B			.360	.661	.200
C				.540	.480
D					.296
E					

1972), and other types of modeling. An ordination of five nematodes at 18 forest sites in Indiana is presented for the XY plane in Figure 4.4. The circle size depicts relative numbers, the letters refer to sites, and the reference points on the axes are arbitrary distances.

In a hierarchical classification, the emphasis is on the extraction of groups at successive levels of relationships. There are two ways to extract

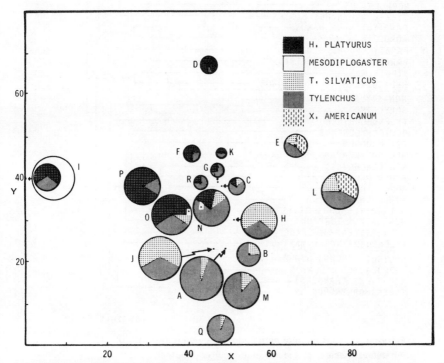

Figure 4.4.   Ordination of nematode communities in 18 forest sites in Tippecanoe County, Indiana. Circle size depicts relative numbers. (From Johnson et al., 1973.)

this information—by subdivisive and agglomerative methods. The subdivisive methods begin with the complete population, and divisions are made successively until a stopping rule is reached. An agglomerative method begins with individuals and combines the individuals most alike until all individuals eventually are united in a cluster. Subdivisive methods

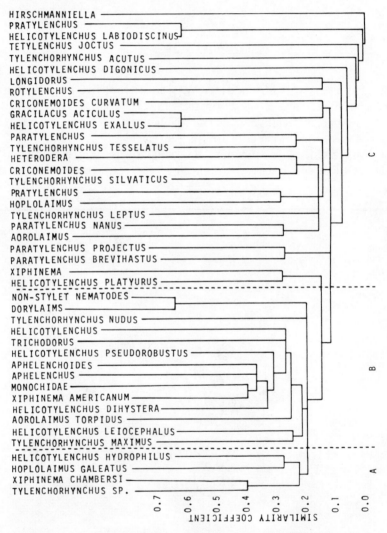

Figure 4.5. Clustering of nematodes in Iowa prairies. A: Pothole area-inhabiting species; B: species that occurred together frequently in well-drained sites; C: nematodes that occurred infrequently but clustered together once with another species. (From Schmitt and Norton, 1972.)

essentially concentrate on differences and thus might not provide natural associations, while agglomerative methods concentrate on similarities.

The subdivisive method was initiated by Williams and Lambert (1959). It consists of subdividing communities until homogeneous groups are obtained. The data are set up in presence-absence form from which 2×2 contingency tables are constructed. Chi-square values are calculated from the contingency tables. Then all chi-square values are summed for each species. Then the species with the largest summed chi-square value is used to divide the location into two groups, depending upon whether or not they possess that species. Within each group the process is repeated, and this goes on until none of the groups contain any significant chi-square values. The end result is a binary classification.

The agglomerative method builds clusters based on similarities within. An example in nematology is that of Schmitt and Norton (1972) (Figure 4.5). There are several ways to do this, but only two will be mentioned. In single linkage clustering, the method first clusters the elements most closely related, gradually admitting more members to the cluster by lowering the requirements for admission. Single linkage means that a high similarity coefficient between a point and any element of the cluster warrants admission of that point to the cluster. Although two clusters may bond, many of the elements of the two clusters may be far removed from each other. Complete linkage clustering is the same as clustering by single linkage, except that the level for admission of a new element to a cluster is a complete linkage level. A given element joining a cluster at a certain similarity coefficient level must have relations at that level or above with every member of the cluster. This has the effect of creating dense clusters compared with single-linkage methods.

Clustering is one form of modeling. Assumptions that are scarcely realistic often are made in constructing models. Models under artificial and sophisticated conditions may serve as a standard against which numerous deviations can be compared, but frequently they cannot display the true situation existing in nature. Then new models are constructed to account for the discrepancies. Even in the computer age, some problems cannot be solved with models, but modeling has its place. In nematology, a number of computerized models such as those of H. Ferris (1976) and Kimpinski et al. (1976) have been used. A computerized model with a fair degree of fitness has been constructed for *Meloidogyne* species and grapes in California (H. Ferris, 1976). Modeling does not have to be complex just because it is the vogue. One should first understand the basics and then gradually develop more complex systems.

POPULATION CHANGE

There seem to be three essential requirements for successful prediction. You need a backlog of full detailed, factual knowledge. You need good reasoning power. And you also need (and this is outside your control) "good behaviour" on the part of the objects whose performance you are predicting.

E. C. Pielou, 1972

Understanding life cycles and the factors governing changes in nematode populations is basic to understanding nematode distribution and plant disease. Even though many empirical data have been gathered, we are just beginning to explore some modern approaches in ecology. Unfortunately, we have not used some of the older methods to full advantage. Of the many books on population change, the reader is referred to those of Andrewartha and Birch (1954), Boughey (1973), Pielou (1974, 1977), and Wilson and Bossert (1971) for introductory and advanced treatments of theory and practice. At our present stage of development, the nematologist must borrow from sources such as these. Some of the pioneering work in population change studies with nematodes has been by F. G. W. Jones, Seinhorst, and Oostenbrink, but other works are scattered in the literature. At the risk of creating misinterpretations, this chapter presents only a simplified introduction to the field. This introduction may encourage researchers not familiar with the subject to delve into it more deeply.

Some people have devoted their lives to population ecology and prediction. Most admit that there is much to learn, and some question if we will ever be able to predict accurately, even with the help of computers. Some say that it might take 20 to 50 years to learn enough about any one pathogen or disease to be confident of prediction success. It is an extremely complex undertaking, but that does not mean that we should not try. It does mean, however, that nematologists properly trained in computer science, statistics, and ecology should assume the lead. Many techniques are still experimental and must be tested over time. Hasty conclusions resulting from inadequate knowledge of the advanced, and sometimes popular, techniques should be avoided.

Practical answers need not always be sought when studying populations. The fundamental question remains, "Why do some organisms increase greatly in a given environment and others do not?" This has been asked many times by ecologists in many disciplines. Some nematode populations seldom attain sufficient density or spatial size to be economically important. Others attain such proportions occasionally and only under rather narrow environmental limitations. Other nematodes attain sufficiently large populations yearly to be a perennial problem. Population measurements usually are made on a single species relative to economic thresholds or injury levels, but collective densities or biomasses of all or most parasitic species might be more important.

When we speak of a population, what do we mean? We often refer to density, that is, the number of individuals per unit of soil. Density often represents a larger spatial population. But how large is the population spatially, and how much extrapolation is justified? When we make a composite sample, have we sampled one population or several discrete ones? Are the seemingly discrete populations actually discrete or are they joined by labyrinths of interconnecting continuums? What is the spatial configuration of a community of nematodes? What is the size of a population? Is it the size of a sample, the size of a plot, or the size of a field? These questions may not be important for many purposes, but experimentalists should be cognizant of them so they can be satisfied that their sampling techniques are compatible with objectives of the investigation. The boundaries of ecosystems, as defined by humans, are usually open.

Sampling is a critical problem in most studies. The importance of samplings is indicated by the many studies that have been made on the subject. Southey (1974) reviewed some of the work on sampling for cyst nematodes. Proctor and Marks (1974) found that frequently used sampling methods are of low precision. Methods of obtaining a relatively high precision in recovery of *Pratylenchus penetrans* were very time consuming. At least 7 hours of sampling and laboratory analysis was necessary to obtain a mean population density per 0.01 hectare plot of the true mean with 95% confidence. In addition, Barker et al. (1969a,b), among others, found that dif-

Table 5.1. Maximum and minimum levels of some plant-parasitic nematodes in the United States and Canada

Nematode	Location	Crop	Maximum	Minimum	Reference
Ditylenchus dipsaci	Utah	Alfalfa	1. Early fall 2. Late spring		Tseng et al., 1968
Tylenchorhynchus claytoni	West Virginia	Red Pine seedbed	Late June	Mid-March	Sutherland and Adams, 1966
T. maximus	Iowa	Prairie	Winter	June–July (dry summer)	Schmitt, 1973a
Pratylenchus brachyurus	Florida	Citrus	June–July	March–May	O'Bannon et al., 1972
P. coffeae	Arkansas	Strawberry	April–May (roots)	Winter (roots)	Riggs et al., 1956
P. penetrans	Ontario	Rye-tobacco	Fall	Summer	Olthof, 1971
P. penetrans	New York	Corn	1. July 2. September	Winter	R. E. Miller et al., 1963
P. penetrans	New Jersey	Strawberry	1. June (soil) July (roots) 2. September (roots)	January	Di Edwardo, 1961
P. penetrans	Indiana	Rotation		Winter	Ferris, 1967
P. penetrans	Wisconsin	Potato	August	February	Dickerson et al., 1964
Radopholus similis	Florida	Citrus	October–December	February–June	DuCharme, 1967

Nematode	Location	Host	Optimal sampling period	Extremes of growing season	Reference
Heterodera rostochiensis	Newfoundland	Potato	Mid-July to early August		Morris, 1971
Meloidogyne graminis	Virginia	Bermuda grass	May	March	Laughlin and Williams, 1971
Rotylenchulus reniformis	Georgia	Cotton, soybeans	August–September	June	Bird et al., 1973
Criconemoides xenoplax	California (Merced Co.)	Grape	Winter and early spring	Late summer and fall	Lownsbery, 1959
Hemicycliophora arenaria	California	Citrus	January	August	Van Gundy et al., 1968
H. zuckermani	Massachusetts	Cranberry	1. July 2. November	Winter	Zuckerman et al., 1964
Tylenchulus semi-penetrans	Florida	Citrus	April–May November–Dec.	August–Sept. February–March	O'Bannon et al., 1972
Xiphinema americanum	South Dakota	Cottonwood	1. June–July 2. Early autumn	February	Malek, 1969
X. americanum	Iowa	Alfalfa	1. July–August 2. March–April	December–February	Norton, 1963
X. americanum	Kentucky	Tobacco sod rotation	1. Spring 2. Fall	1. Winter 2. Summer	Flores and Chapman, 1968
X. americanum	Wisconsin	Blue spruce	1. June 2. Fall	1. February 2. Late summer	Griffin and Darling, 1964

ferent population patterns were obtained with different methods of extraction and storage.

A major goal of agricultural nematology is to prevent nematode populations from increasing to economic threshold levels. Eradication is seldom attained and it usually is not desirable for economic reasons if nothing else. Also it has never been demonstrated that plants growing in a nematode-free environment are any better off than those with few parasites. Pathogens often are unnoticed because they are not sufficiently abundant to devastate a crop. Injury may be present; it is just not obvious. From a pragmatic standpoint, for control we are not so interested in presence-absence, unless it reflects potential, as we are in maintaining the mono- or polyspecific population below the injury level.

SEASONAL PATTERNS

In many respects, nematodes are no different from other organisms in that their populations fluctuate during the year and in different environments. As expected, population fluctuations vary with the geographical location (Table 5.1). The least recovery of nematodes in cultivated fields is often in winter or early spring in the northern latitudes, where cold temperatures probably are limiting. In parts of California and in Florida, the period of low recovery often occurs in summer or early fall when hot, dry conditions exist or when the crop is not in lush growth. Primary and secondary peaks often occur, and they vary considerably. Fluctuations within the main growth curve occur during a growing season and often are correlated with major divergences in temperature and moisture. Thomas (Figure 4.1) demonstrated that nematode densities vary within small areas and among species in corn under different tillage systems. Likewise, densities of *Helicotylenchus hydrophilus, Merlinius joctus,* and *Tylenchorhynchus* sp. during a season varied considerably within a few meters of each other in a native prairie (Schmitt, 1973a). Pitcher and McNamara (1970), however, found that *Trichodorus viruliferus* did not exhibit major fluctuations but reproduced all year even when the soil temperature was 5 C. These authors concluded that the food supply was the determining factor in reproduction, at least compared with temperature. Populations may differ in a season depending on the degree of host susceptibility as demonstrated by A. W. Johnson et al. (1974) and Minton et al. (1960). In the latter study, juveniles of *Meloidogyne incognita acrita* increased more slowly in plots containing resistant cotton plants, such as *Gossypium barbadense* and Auburn 56, than under the susceptible varieties Empire and Rowden (Figure 5.1). Nematodes other than *Meloidogyne* responded differently throughout the season on different cottons in a naturally infested field. Several studies indicate that not

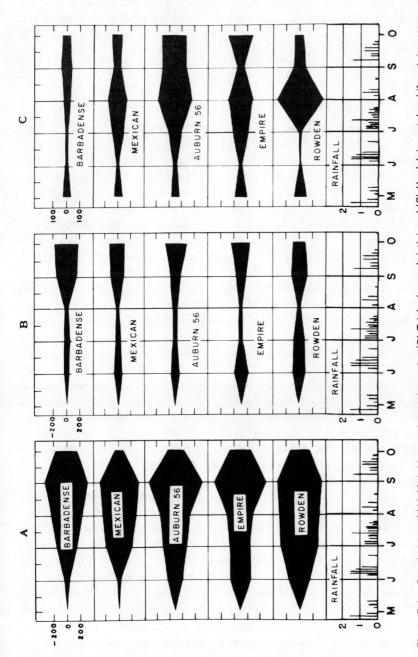

Figure 5.1. Populations of (A) *Meloidogyne incognita acrita,* (B) *Trichodorus christiei,* and (C) *Hoplolaimus tylenchiformis* in cotton plots. (From Minton et al., 1960.)

85

only are densities frequently smaller at deeper depths, but they fluctuate less as sampling depths increase. Females of *Xiphinema americanum* contained seasonal peaks at all levels, but these became less evident with juveniles at progressively greater depths (Griffin and Darling, 1964).

POPULATION CHANGE

Many factors determine the rate of increase and eventual size of a population. Among these are fecundity, fertility, duration of life cycle, longevity, substrate available for parasitism, and the physical, chemical, and biological environment. Since many parameters are interrelated, probably no one factor solely governs nematode growth, but one factor can be limiting. Nematode densities can change suddenly and should be studied over time. Eggs are a 'part of the population, but they usually are not studied because of extraction difficulties and sometimes because of the even greater difficulty of distinguishing an egg of one species from that of another. Numerical data of relatively easily obtainable eggs such as those of *Heterodera* and *Meloidogyne* have been used for some time, however. Galls and other symptoms sometimes reflect population size.

A susceptible host is one of the most important factors governing populations of obligate parasites, as most plant nematodes seem to be. Any increase from internal food reserves would be minimal and ephemeral without a host. A resistant plant might maintain a population and serve as a time bridge until a more susceptible host becomes available. A tolerant host could not only act as a time bridge, but could also actually allow an increase in density or spatial size of a population. Since an objective in most agricultural areas is to maintain nematodes at low levels, knowledge of population curves that occur in different environments is important. It is easier to measure the changes in population density, however, than it is to understand the functional dynamics of these changes.

It is well known among animal ecologists and naturalists that generally there is an abundance of food relative to animal density (Andrewartha and Birch, 1954; Cockerell, 1934; Darwin, 1859). When one considers the thousands of species capable of attacking plants, the number of species attacking a given crop in a given field narrows to a relative few. Furthermore, with nematodes as with other organisms, a plant usually can support more individuals than it actually does. Rarity of parasites is the rule and not the exception, relatively speaking. Food may be a limiting factor in population growth, but environmental resistance often is more important than is food availability in restricting nematodes. Few populations remain constant in size or density, a state of equilibrium being reached only when immigration equals emigration and when mortality equals natality. Although the former are important in population changes of many animals, natality and mortality doubtless account for the major changes in populations of plant-parasitic nematodes. A population or community in its early developmental stages usually is small in biomass, lacks stability, possesses a high production/respiration ratio, and has a high net production. In its mature stages it usually has a large biomass, a low net production with more energy flow going toward maintenance than net gain, and greater stability.

Fecundity and Fertility

Because not all potential reproductive units are viable, fecundity and fertility are not synonymous. Fecundity reflects the number of potential propagules produced, be they eggs, spores, or any other means of multiplication, whereas fertility is the number of viable units produced. *Ditylenchus dipsaci* (Yuksel, 1960) and *Radopholus similis* (Loos, 1962) curtail egg production shortly after being placed into water. This suggests that nutrition, aeration, or other environmental factors can affect fecundity. Fisher (1969) found that a continuous food supply was required for *Aphelenchus avenae* to produce eggs. Under adverse conditions of temperature, and with unfavorable qualitative and quantitative nutritional factors, eggs per female were produced more slowly and over a longer time than under more favorable conditions, but the total laid was about the same under both conditions.

The largest egg counts usually have been obtained with species of *Meloidogyne, Heterodera,* and *Rotylenchulus,* where eggs are contained in an egg sac or cyst or, as with species of *Anguina, Ditylenchus,* or *Aphelenchoides,* where the nematodes are inside a gall or enclosed in plant tissue. Do other types such as *Pratylenchus, Helicotylenchus,* or *Tylenchorhynchus* lay just as many, which are not recorded because they are not contained? Although little is known concerning fecundity and fertility of most nematodes, we can make educated guesses as to their magnitude by the abundance of individuals. A species that is not prolific would not be abundant unless it has superior survival capacities. Some data are presented in Table 5.2 for number of eggs produced and in Table 5.3 for population increase. Most of these data are from greenhouse or laboratory tests and are therefore probably representative of favorable conditions. Figures for field populations can be expected to be much smaller in many instances. Such data provide information on the capacity of a nematode to increase. The fact that nematodes usually do not attain this capacity in the field is indicative of environmental resistance. This can be exemplified by species of *Seinura.* From the rapidity of their life cycle, one might suppose that these

Table 5.2. Number of nematode eggs produced

Nematode	Eggs per individual	Average	Comments	Reference
Aphelenchoides ritzemabosi		2/day		French and Barraclough, 1961
Aphelenchus avenae		207	25 C for 37 days	Fisher, 1969
Ditylenchus dipsaci	From 207 to 498 eggs in 45–73 days		Onion seedlings	Yuksel, 1960
Meloidogyne sp.	High of 2882 by one female; above 500 was common	24–121/day		Tyler, 1938
Paratylenchus projectus		3/day	Red clover in agar	Rhoades and Linford, 1961b
Pratylenchus penetrans	68 maximum	1.1/day	Varied with temperature	Mamiya, 1971
Radopholus similis		1.8/day	In water; production declined quickly after 24 hours	DuCharme and Price, 1966
R. similis	2.6/female			Loos, 1962
Rotylenchulus reniformis	Maximum of 196 per egg mass	121	Eggs plus empty shells	Linford and Oliveira, 1940

would be among the most abundant of nematodes. But they are not especially common in nature, even where potential prey seem abundant.

Natality and Mortality

Natality is the rate of production of new individuals. It is usually divided into ecological natality, the actual rate of production of new individuals in a specific environment, and maximum or absolute natality, the rate that would obtain if there were no environmental resistance. Absolute natality is limited only by the inherent physiology of the nematode and is a constant for each species, but a constant that we can only surmise. In nature, absolute natality is probably seldom obtained except for short durations. Because the data in Table 5.3 are mostly from tests conducted in the greenhouse, the figures are probably closer to absolute natality than to ecological field natality.

Natality can be expressed as

$$\Delta N_n = \text{natality}$$

$$\frac{\Delta N_n}{t} = \text{natality rate per unit of time.}$$

$$\frac{\Delta N_n}{\Delta t / N} \quad \text{or} \quad \frac{\Delta N_n}{N \Delta t} = \text{natality rate per individual per unit of time}$$

The figures of nematode increases in Table 5.3 might reflect the growth rate more than the natality rate, however, because natality represents the number of new individuals and not the net increase. The two are not the same because mortality may be considerable. The growth rate may either increase or decrease, but the natality rate is never less than zero. It is difficult to obtain natality rates for nematodes in nature because we do not know the longevity of nematodes in field environments. Occasionally we can deduce that some nematodes live for months and possibly over a year in the field. It is also possible that some individuals live only hours to days after hatching. Therefore, in most studies, population growth rates are probably more useful than natality figures, although the latter would indicate the potential increase if it could be realized.

Mortality is difficult to measure with nematodes because of the generally rapid decay of most individuals following death. If natality and longevity of individuals are known, mortality would be reflected in the population size. If a population is declining, its mortality is greater than its natality. Minimum mortality results when nematodes die of old age. Since nematodes are killed more easily at some stages of development than at

Table 5.3. Population increase of some nematodes

Nematode	Increase	Time	Conditions	Reference
Belonolaimus longicaudatus	From 350 to ~15,000	155 days	Varied with variety and other nematodes	Johnson, 1970
Criconemoides curvatum	$100\times$ from 1000	90 days		Malek and Jenkins, 1964
C. lobatum	From 900 to 100,000	7 months	Grass	Johnson and Powell, 1968
C. ornatus	From 1000 to ~113,000	155 days	Bermudagrass, varied with variety and other nematodes	Johnson, 1970
C. xenoplax	From 100 to 12,000	3 months		Sher, 1959
Hemicycliophora parvana	From 200 to 17,025	5 months	Carnation	Ruehle and Christie, 1958
H. similis	$34\times$	90 days	Cranberry	Bird and Jenkins, 1964
Hoplolaimus galeatus	From 200 to 8102	13 months	*Pinus ponderosa*	Riffle, 1970b
Paratylenchus dianthus	From 100 to 1,509,000	105 days	Celery; varied with host	Rhoades and Linford, 1961b
P. hamatus	From 2000 to 106,048	76 days	Peppermint; varied with host	Faulkner, 1964

Species	Number	Time	Host/condition	Reference
P. projectus	From 100 to 2,637,000	105 days	*Nicotiana alata* var. *grandiflora*, varied with host	Rhoades and Linford, 1961b
Rotylenchus pumilis	From 200 to 3694	13 months	*Pinus ponderosa*	Riffle, 1970b
Scutellonema brachyurus	692× from 90	325 days	Red clover	Chapman, 1963
Tylenchorhynchus claytoni	From 19 to 30,000	13–15 weeks	23 C	Khera and Zuckerman, 1962
T. cylindricus	From 200 to 2088	13 months	*Pinus ponderosa*	Riffle, 1970b
T. martini	From 1800 to about 98,000	155 days	Bermudagrass, varied with variety and other nematodes	Johnson, 1970
Tylenchus agricola	From 15–17 to 27,000–30,750	3 months	23 C alfalfa callus	Khera and Zuckerman, 1962
Xiphinema americanum	From 200 to 1359	13 months	*Pinus ponderosa*	Riffle, 1970b
X. diversicaudatum	From 200 to 788	13 months	*Pinus ponderosa*	Riffle, 1970b
X. diversicaudatum	From 200 to 27,000	39 months	Rose	Schindler, 1957

Table 5.4. Time for completion of life cycle of some nematodes

Nematode	Time	Remarks	Reference
Aphelenchoides besseyi	9–11 days	23–30 C on *Fusarium solani*	Huang et al., 1972
A. ritzemabosi	11–12 days	Mean temperature between 17 and 23 C	French and Barraclough, 1961
A. ritzemabosi	13–14 days	Mean temperature between 13 and 18 C	French and Barraclough, 1961
Criconemoides xenoplax	25–34 days		Seshadri, 1964
Ditylenchus triformis	16–21 days	Egg to egg, 24–26 C	Hirschmann, 1962
Helicotylenchus pseudorobustus	30 days; 35 days	32 C; 24 C	Taylor, 1961
Heterodera betula	52 days	28 C, but reproduction occurred down to 14 C	Riggs et al., 1969
H. lespedezae	25 days	26 C	Bhatti et al., 1972
H. rostochiensis	One generation/year	Newfoundland	Morris, 1971
Longidorus africanus	Less than 4 months	20–23 C	Cohn and Mordechai, 1969
L. elongatus	19 weeks		Yassin, 1969
L. elongatus	9 weeks	Egg laying to adult, 30 C	Wyss, 1970a
Meloidogyne naasi	39–51 days	Wheat, 26 C day, 29 C night	Siddiqui and Taylor, 1970b
Merlinius joctus	32–36 days	30 C	Norgren et al., 1968
	36–40 days	22 C	

Species	Duration	Conditions	Reference
Paratylenchus projectus	23 days	Preadult to preadult	Rhoades and Linford, 1961b
Pratylenchus neglectus	28 days		Mountain, 1954
P. penetrans	86 days	15 C	Mamiya, 1971
	35 days	24 C	
P. penetrans	35 days	Potato or onion	Wong and Ferris, 1968
Radopholus similis	18–20 days	24–27 C	DuCharme and Price, 1966
Rotylenchulus parvus	27–36 days	24–28 C	Dasgupta and Raski, 1968
R. reniformis	17–23 days		Birchfield, 1962
Seinura celeris	2 1/2 days	28 C	Hechler and Taylor, 1966
S. demani	4 1/2–5 days	20 C	Wood, 1974
S. oliveirae	2 1/2–3 days		Hechler and Taylor, 1966
S. oxura	3 1/2 days		Hechler and Taylor, 1966
S. steineri	6–6 1/2 days	28 C	Hechler and Taylor, 1966
S. tenuicaudata	5 1/2–6 days	28 C	Hechler, 1963
Trichodorus viruliferus	45 days	In laboratory tubes	Pitcher and McNamara, 1970
Tylenchus emarginatus	5–6 days at 25 C; 6–7 days at 20 C; 6–7 days at 17–23 C	Sitka spruce on agar; egg to egg; no reproduction found at 30 C or below 10 C	Gowen, 1970
Xiphinema americanum	At least 1 year	South Dakota	Malek, 1969
X. index	22–27 days	Oviposition to adult	Radewald and Raski, 1962a

others, even small judicious manipulation of the environment may be important in nematode control.

Population Increase

Being consumers nematodes depend on a ready food source, which in turn will govern their populations. An initially small population doubtless lives in a precarious balance since a major adverse disturbance may eliminate it. After a population is established and stabilized, however, a disturbance of the same magnitude might be detrimental to the population but not fatal. Because nematode reproduction is slow compared with many microorganisms, a sudden increase in population size is apt to be the result of a sudden hatching of eggs. A sudden decrease, however, is probably due to a sudden adverse environmental change and not to an inherent curtailment of egg laying.

The known life cycle of nematodes usually varies from a few days to a year or more (Table 5.4). Temperature is important in regulating the rapidity with which the life cycle progresses. Cold can increase the length of the life cycle of nematodes by putting them in a physiologically quiescent state. The mean length of adult life of *Tylenchus emarginatus* feeding on Sitka spruce in agar was 128 days at 15 C and 32.5 days at 25 C (Gowen, 1970). The total number of viable eggs produced at 15 and 25 C was not affected, however. Mamiya (1971) found that even though the optimum temperature for reproduction of *Pratylenchus penetrans* was 24 C, closely followed by 30 C, fecundity was about the same at 15, 20, and 24 C, but it was suppressed slightly at 30 C and inhibited at 33 C.

GROWTH PATTERNS

Our goal in economic nematology is to limit populations of nematodes to a level below economic injury. Thus an interest in growth curves comes early in a study of any species. Many growth curves have been constructed for other animals, and the basic curves can be applied to nematodes (Oostenbrink, 1966, 1971; Seinhorst, 1966, 1967a,b,c, 1970). The type of growth curve obtained depends on host suitability, the reproductive capacity and reproduction cycle of the nematode, and the environmental conditions. Many growth curves for nematodes have resulted from experiments in the greenhouse or growth chamber where extraneous factors can be minimized. Since nematodes in the field must contend with many parameters, one must discern between the maximum inherent growth (biotic potential) of a nematode and growth imposed by the environment (ecological growth). The biotic potential is largely theoretical, but it can be attained if all conditions are favorable. Since these conditions often are unknown, we must rely on the maximum growth rate obtained as being the closest thing to the true biotic potential. The difference between the biotic potential and the ecological growth curve is due to environmental resistance.

Exponential Growth

There are two basic types of growth curves, each with several variations, of which only the simplest will be considered here. The first is the exponential growth curve where the reproduction rate is constant. When there is no overlap of generations and survival is only by eggs, the equation

$$N = R_0^t N_0$$

can be used, where R_0 is the replacement for each nematode, t is the number of generations, and N_0 is the initial population at the start of the observations. Let us assume that a nematode is living in a habitat where there is only one generation per year and that only eggs overwinter. We will also assume that we start with 60 females and that every female in one generation will produce a net replacement of 2.7 females (we will assume also that it is a parthenogenetic species). If growth is exponential and we wish to predict the number of nematodes after 6 years, the calculations would be

$$N = 2.7^6 \times 60$$

or 23,245 nematodes after 6 years.

An equation commonly used for exponential growth where reproduction is continuous is

$$\frac{dN}{dt} = rN$$

where N is the number of individuals at a given moment, t is the unit of time, and r is the intrinsic growth rate per unit of time. The birth rate minus the death rate is represented by r, which, although it will vary, is used for our purposes as if it were constant. For example, if a nematode population increased from 360 to 862 in a month, r would equal $(862 - 360)/360 = 1.39$ per month. Since we started with 360 nematodes , the rate of population increase equals $1.39 \times 360 = 500$ nematodes per month. However, perhaps due to different age structures in the population, r is not always constant. A derivative of $dN/dt = rN$ results in the equation $N = N_0 e^{rt}$, which can be used to predict N for any time desired. N_0 is the initial population at any given starting time, e is the constant 2.71828, which is the base of the natural system of logarithms, r is the intrinsic growth rate, and t is the elapsed time. As an example, if the present population of a given nematode is 27 nematodes per unit of soil, and r is 0.79/month, what would the population be in 12 months? The calculations are:

$$N = N_0 e^{rt}$$

$$N = 27 \times 2.71828^{0.79(12)}$$

$$N = 353{,}568 \text{ nematodes per unit of soil in 12 months}$$

Exponential growth results in a J-shaped curve, and if it should continue it would not be long before the total nematode mass would be expanding at a rate close to the speed of light. Fortunately that does not happen. But there are many examples of exponential growth of nematode populations, for short periods, in both the field and the greenhouse. Exponential growth often occurs when populations are small and growth is unrestricted. Sooner or later the population growth rate is reduced or even reversed by an adverse environment, which may include density-dependent factors. This brings us to the logistic growth curve.

Logistic Growth

As we have seen, a population does not increase exponentially forever, or for long. As the nematode numbers increase, there comes a time when there may be too many individuals for the environment to sustain the populations exponentially, and the increase begins to slow. Or the environment may change suddenly, bringing reproduction to an abrupt halt. When the number of new individuals equals the number of dying, as supported by environmental conditions, then the carrying capacity, usually designated as K, or the asymptote, is reached. The simplest equation for logistic growth is

$$\frac{dN}{dt} = rN \frac{K - N}{K}$$

This merely means that as N increases, any change, i.e., dN/dt, will decrease. As N approaches K, dN/dt will approach zero. When N is small, assuming the environment is favorable for increase, then the growth rate will be exponential or nearly so, at least in the early stages of growth. Because of the Malthusian principle, in which a population can temporarily breed beyond its food supply, the growth curve may "overshoot" the carrying capacity. After the asymptote, sometimes called the P_{max}, is reached, the population will either stabilize, somtimes in a fluctuating manner, or decline. The decline may be slow or rapid. The curve obtained by Faulkner (1964) using *Paratylenchus hamatus* illustrates one type of logistic curve (Figure 5.2). The curve is largely exponential until about the twelfth week,

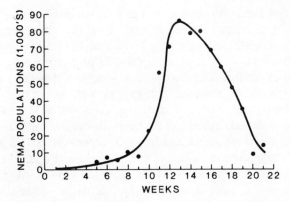

Figure 5.2. Population curve of *Paratylenchus hamatus* on peppermint during 22 weeks. (After Faulkner, 1964.)

at which time the peak is approached. The maximum population is reached at the fourteenth week and then declines sharply.

Many equation for population growth have been proposed, but most things can be condensed to simple basics. Oostenbrink (1966) put it well:

> The socalled logistic equation, as a quantitative expression of population growth, is published and republished in different notations as a more or less original finding by several authors. All these seemingly complicated formulae about population increase comprise only two simple elements, already expressed by R. N. Chapman (1928) and reformulated more exactly by later authors, namely that the number of animals of a species is determined by its "biotic potential = a potential or geometric rate of increase" and by "environmental resistance = reduction of this potential rate of increase by a factor proportionate to the unrealized part of the potential increase."

Populations Relative to Available Substrate

The amount of food available to a nematode is affected by the number of nematodes feeding on a given site, with a consequent increase or decrease in the nematode population. Competition for food among species may shift their relative numbers. There may be indirect effects of density-dependent factors from parasitism of the host. For example, the parasitism of a plant by *Meloidogyne* sp. may not depend on the density of the same or other species of nematodes. Consequences of parasitism, however, through galling and other morphological and physiological changes could restrict feedings by other nematodes.

Nusbaum and Ferris (1973) analyzed population increase relative to cropping systems. Population increase depends upon the initial population (P_i), the maximum reproduction rate on a given crop (a), and the proportion of noninfested tissue ($1 - x$), where x equals the proportion already infected. The objective of control is to maximize $1 - x$. This depends partly on the rapidity of growth of susceptible tissue as well as on the amount of parasitism on the plant. At low P_i densities, $1 - x$ will be large at the beginning of the season and, other factors being favorable, the plant will grow vigorously. As the plant grows and fruits, a larger root system is needed to support the foliage. The net damage done will depend upon the size of the nematode population attained relative to the plant growth. At the beginning of the season, $x = 0$, since there is no infected tissue. When the nematode increase is just sufficient to maintain the population density, then equilibrium density (E) is attained.

Naturally, the susceptibility of a host and the preplant density will affect the size of the nematode population. Seinhorst (1967b) has summarized ways that nematode populations can be affected (Figure 5.3). The dark straight line where any initial nematode density (P_i) equals the final density (P_f) is the maintenance line. Points above the maintenance line indicate that the population at a given initial density has increased while points below the

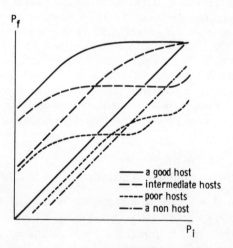

Figure 5.3. The relation between initial and final densities in experiments with a nematode on a good host, intermediate hosts, poor hosts, and a nonhost. P_i and P_f: initial and final densities on logarithmic scales. (From Seinhorst, 1967b, p. 431.)

Table 5.5. Reproduction of *Pratylenchus vulnus* and its influence on growth of rose after 5 months (modified from Santo and Lear, 1976)

Treatment (P_i)	Fresh plant weight (g)	P_f (1000)	Specific growth rate
Control	49.4	0	0
100	34.2	333	3,329
1,000	26.3	398	397
10,000	17.4	212	20

line show that the population has decreased. To minimize the effects of nematodes in any crop management system, the susceptibilities of the current crop and succeeding crop should be known, as well as the nematode's ability to survive. The growth rate of a nematode is not always positively correlated with the P_i number. In fact, they are often inversely related owing to the poorer growth of the plants at higher P_i numbers. For example, Santo and Lear (1976) found that the P_f of *Pratylenchus vulnus* was inversely proportional to the P_i (Table 5.5).

Underpopulation

Seinhorst (1968) states, "A habitat is underpopulated with a species if at higher density this species would increase in number at a faster rate." He states further that underpopulation can occur only in amphimictic species in which the multiplication rate depends on the meeting chances of males and females.

In an experiment with *Rotylenchus uniformis*, the rate of multiplication of peas in pots was greater at the higher initial densities than at lower ones, but the proportion of adults to juveniles was lower at higher initial densities than at lower ones. Seinhorst contends that this supports the hypothesis that at very low densities the proportion of adult females that mated was smaller than at higher initial densities because the chance of a female meeting a male decreased with density below a certain density. Thus, the proportions of adults in the final populations were much larger at lower initial densities. Seinhorst explained this shift as follows: At low densities, the nematodes fed and developed to adults on the roots, but many fewer offspring were produced per female than at higher densities. The smaller number of progeny at the lower densities could, therefore, only be due to a smaller proportion of the females having become fertilized than at higher densities.

r and *K* Selection

Two of the main components of the logistic curve are *r*, the intrinsic growth rate, and *K*, the asymptote. Much has been made of *r* and *K* selection (strategy), but, as Pielou (1975) has pointed out, their wide applicability remains to be discovered. The terms are widely used, however, and they may have interpretive meaning in polyspecific communities. One of the characteristics of the *r* selection nematodes is their ability to exploit ephemeral resources rapidly. The *r* selection species are generally thought to have little competitive ability. They are the species that first exploit a substrate. The organisms best suited for this usually are relatively small species with a high reproductive rate and usually have a short life span. Species with a low reproductive rate could not compete effectively for temporary substrate such as occurs in annual crops. Although the abiity to move about quickly and find new substrate is important to *r* selection species in some animals, it is minor with plant-parasitic nematodes unless the habitat is scaled down to microenvironments. It is also thought that *r* selection species, because of the temporary nature of the substrate, are kept on the lower part of the logistic curve.

The *K* selection species are better competitors in the long run and live in a more stable habitat than *r* selection species. They have larger bodies, lower reproductive rates, longer life spans, and sometimes greater specialization. Thus their intrinsic growth rate is less, but they have greater competitive ability and operate near saturation for the environment in which they occur. Because both types are results of evolutionary processes, they fill different niches. But not all nematodes fit neatly into one or the other of the two basic categories; gradations occur.

How do these terms relate to populations of plant-parasitic nematodes? Caution must be employed, as it is so easy to use various examples because they fit the theory and conveniently ignore those that do not. But keeping this in mind, one can speculate until more data are obtained. It was mentioned in Chapter 3 that species of *Pratylenchus* were not common in native prairies and woodlands. Yet when field crops are grown on land converted from natural areas, individuals of *Pratylenchus* are perhaps the most common parasitic nematodes. Could it be that the unstable habitat of cultivated fields, plowed annually, favor their *r* selection capabilities and that they are not competitive in the more stable habitats of the prairies? Or is it a matter of host preference, or have edaphic factors been changed so much by cultivation that this itself favors the nematode? In many ways, species of *Pratylenchus* fit the *r* selection group. They are small, the life cycle of many species is about 30 to 45 days (Table 5.4), shorter than that of many of the larger nematodes, and populations are often high in annual crops. But the situation is not clear in *Criconemoides* and related genera. These nematodes are generally small and often multiply rapidly. Yet they are more common and frequently in much greater numbers in the more stable habitats than in cultivated soil, although there are exceptions. Are these *r* or *K* species? Or has some type of specialization resulted in their thriving in stable habitats in spite of their small size and biotic potential? Johnson et al. (1974) consider the dorylaimids as extreme *K* strategists in forest habitats in Indiana, and the rhabitids as extreme *r* strategists. The tylenchids as a group were considered neither extreme *r* nor *K* strategists, as some species do well in stable habitats whereas other readily utilize disturbed areas. Each habitat probably should be considered individually. We need to accumulate many data and analyze them correctly before we can put the *r* and *K* selection species of nematodes into perspective.

NEMATODE DENSITIES AND SEX RATIOS

The reproductive capacity of nematodes is intergral with population size, and any change in numbers of a given sex or in sex ratios would potentially alter the resulting numbers of progeny. The environment is important in causing shifts in sex ratios. However, little is known about how these shifts affect long-term population growth, although conceivably the effects could be drastic. Several environments or factors are associated with sex ratio changes, and in most instances it is not known whether the effect is direct or indirect—for example, through physiological alteration of the host with subsequent morphological changes.

Host nutrition is important in altering sex ratios, usually with a shift towards males as nutrition becomes more unfavorable for plant growth or when less food in the form of host tissue is available per nematode. Tyler (1933b) found that 0.7% of *Meloidogyne* juveniles became males in single nematode cultures, but that males were more numerous in old, unhealthy, or heavily parasitized roots. In multiple primary infections, 16.4% of the adults were males, while in multiple secondary infections, 56.5% developed into males. Triantaphyllou (1960) reported that, when few juveniles of *M. incognita* entered a tomato root, most became females, but when crowding resulted by invasion of many juveniles, most became males. Similar results were obtained with *Heterodera rostochiensis* by Trudgill (1967), who found that removal of plant tops, which probably reduced root carbohydrates, increased the male/female ratio, while the total number of nematodes was not altered. Trudgill believed that only second-stage juveniles with giant cells of sufficient size and nutrition became females. He also found that most juveniles that invaded lateral roots became males. Similarly, Ellenby (1954) found a greater proportion of males in lateral roots than in primary roots. Increases in males associated with depletion of nutrients have also been found with *Meloidogyne* spp. (McClure and Viglierchio, 1966; Davide and Triantaphyllou, 1967a,b; Bird, 1960) and with *H. schachtii* (Kerstan, 1969). Kerstan believed, however, that sex was determined before the nematode penetrated the root and that the changes in sex ratio were caused by selective deaths in female juveniles. Ross and Trudgill (1969) concluded that juveniles of *H. rostochiensis* randomly invaded primary roots and that competition for space to produce a large enough group of giant cells was the chief factor determining the sex ratio; the more space available, the more females produced. Thus, the greater occurrence of males in the lateral than in the main roots, as found by others, is possibly due to the lateral roots being thinner and less able to produce groups of giant cells large enough to produce females. Bird (1970) found males of *M. javanica* in treatments in which the juvenile inoculum level was high. He speculated that the growth rate of *M. javanica* is related to the degree of stress to which it is subjected. There is an acceleration of growth rate with a little stress, but increasing stress leads to production of males and a slowing of the overall nematode growth rate.

Ketudat (1969) demonstrated that fungi may be involved in nematode sex ratios. He found that the male/female ratio of *H. rostochiensis* in tomato increased when *Rhizoctonia solani, Verticillium alboatrum*, or a grey sterile fungus was added. The male/female ratio increased with an increase in the amount of fungus inoculum added. Such phenomena could result from competition for root space; as root necrosis occurs, there is less tissue on which the nematode can feed. Plants infected with *V. alboatrum* supported more males than did plants infected with one of the other two fungi. Wilt symptoms developed more slowly than necrosis caused by the other two fungi. Since there is little root necrosis in the early stages of wilt development, it is possible that more juveniles invaded roots of wilt-infected plants, resulting in more crowding and thus more males.

Although most work with altered sex ratios resulting from depletion of nutrients from overcrowding or other causes has been with species of *Heterodera* or *Meloidogyne*, there is evidence that in at least one instance crowding resulted in a higher female/male ratio. When few nematodes of *Anguina agropyronifloris* were in the floret primordia of western wheatgrass, the adult sex ratio was approximately 1:1 with no more than two adults of either sex within a floret. When many nematodes were in a floret, however, females predominated, with a maximum of 73 females and 8 males being found in any infected floret (Norton, 1965a).

Along with nutrition, the temperature effect on sex ratios has probably been investigated most. Yet a confusing picture sometimes results. Using Lahontan alfalfa, Griffin (1969) found that more juveniles of *M. hapla* developed into males at 30 C than at 15 to 25 C, but this did not hold true for another variety of alfafa. In another study with *M. incognita* on tomato, a significantly higher proportion of males occurred at 15 C than at higher temperatures, but the male/female ratio was low in all instances (Davide and Triantaphyllou, 1967a). When populations of *M. graminis* were exposed to 27 C for 1 to 5 days and then transferred to 32 C, the percentage of males ranged from zero for one day exposure at 27 C to 45.5% after 5 days exposure. Exposure at 32 C for 1 to 2 days, however, resulted in only 1 to 2% males when the population was transferred to 27 C. When juveniles were exposed to 32 C for 3 or more days and then transferred to 27 C, 90% males resulted (Laughlin et al., 1969). Patterson and Bergeson (1967) found that more males of *Pratylenchus penetrans* migrated from roots at 15 C than at 30 C, but there is a possibility that a greater number of males were produced at the lower temperature.

It was found for *Pratylenchus alleni* that moisture can also affect sex ratios (Norton and Burns, 1971). The proportion of males to females was closer to a 1:1 ratio in the dry regime, field capacity to 50% field capacity, than in the wet regime, field capacity to 125% field capacity, where the females greatly outnumbered the males. It is evident that the dry regime favored either male production or survival.

The work of Griffin (1969), in which sex ratios of *M. hapla* were not the same in two varieties of alfalfa, suggests that genetic makeup of the host may also be a factor in the proportion of males or females produced. In this work, a greater percentage of males was found in the resistant variety than in the susceptible variety. Wong and Ferris (1968) found that the ratio of females to total adults was 10% higher in nematodes extracted from potato roots than in those from onion roots. Males of *Heterodera trifolii* have also been found to occur in some lines of white clover and not others (Norton, 1967). Different races of *H. schachtii* resulted in different sex ratios when inoculated on a common host (Steele, 1975). Although these results may seem to be caused by genetic differences, they may reflect nutritional changes.

Other factors have been shown to alter sex ratios in nematodes. Gamma ray irradiation of genital primordia of *M. incognita* prior to sex differentiation resulted in a percentage increase of males (Ishibashi, 1965). Spraying plants with maleic hydrazide resulted in an increase in the percentage of males in *M. incognita* and *M. javanica,* and the earlier the application the greater the percentage of males. Intersexes of *M. javanica* were also found from plants sprayed with this chemical (Davide and Triantaphyllou, 1968). Morphactin caused sex reversal in *M. javanica* in which the percentage of males in the population increased to 60% (Orion and Minz, 1971). The chemical drastically altered the development of syncytia and galls, and thus the effect on the nematode may be indirect. If sex ratios of nematodes could be consistently controlled by chemicals and without phytotoxicity, this would have great practical significance.

Intersexes are known in a few genera of soil or plant-parasitic nematodes (Hirschmann and Sasser, 1955; Steiner, 1923; Chitwood, 1949), but not enough is known about how effective they are in changing the nematode populations in nature. It is suggested by the work of Evans and Fisher (1969) that the abundance of males may adversely affect the overall population. They found that the dominance of males of *Ditylenchus myceliophagus* coincided with the time that nematodes reached the plateau phase of population.

Sex ratios, of course, can be changed by the fecundity of the nematodes and the meeting and insemination of the female by the male. Some interesting work on attraction has been done by Green et al. (1970), who found that females of *H. rostochiensis* and *H. schachtii* attracted males and that one male could fertilize several females. They also found that one female may also be inseminated by several males. Males of *H. rostochiensis* were attracted by both males and females, but they were attracted more strongly by the latter. Females in groups attracted males more strongly than did single individuals. Thus, clustering may be advantageous to a species, especially when a sex ratio stongly favors females. The ability of a single male to mate with several females is also an advantage to such species as *H. rostochiensis* and *H. schachtii*, in which sex ratios vary considerably. Much of the work was done with nematodes on agar and this phenomenon may not be so active in nature. The authors found that insemination of *H. rostochiensis* occurred more readily in agar and peat than in sand. In other work, females of *H. rostochiensis* secreted a sex attractant that was active to 15 cm in sand. All females were mated within 50 days of root invasion by the juveniles, and some within 24 days. Females usually mated soon after they ruptured from the root cortex (K. Evans, 1970).

CHAPTER 6

THE SOIL ENVIRONMENT

Living organisms have hereditary processes which limit or define their environmental requirements.

Soil is the home of most nematodes and deserves more attention than it usually receives. Errors in, or less than adequate usage of, terminology relative to soil and nematodes abound in the literature. Often statements are accepted uncritically and can result in misleading interpretations. This is as disconcerting to the ecologist as an erroneous reference to a taxon is to the taxonomist. For example, what is a clayey soil? Is it clay, clay-loam, silty-clay, silty-clay-loam, sandy-clay, or a sandy-clay-loam? More attention to basic soil principles may help alleviate some of the more serious errors. It is with this thought that this chapter is written, but it is not intended to be a substitute for a course in soils or a good textbook on the subject. Much is yet to be learned about soils by soil scientists, and it will be well worth the effort to integrate their knowledge with information on nematode populations.

Soil is a complex system, generally considered to consist of mineral, organic, gaseous, and aqueous phases with the mineral portion forming the bulk of the matrix. Properties can vary considerably in localities of close proximity. Were the soil not complex, there would be more instances of a few nematode species increasing to large populations instead of, as is frequently the case, several species increasing only slightly or moderately. Some nematodes increase to enormous numbers when placed in a near-optimal physicochemical environment in the greenhouse where there is little, if any, competition from other organisms. Yet these nematodes seldom attain such populations in nature even when comparable root systems are compared. Environmental resistance is operative and one must frequently look to the soil for the answer.

Before discussing specific factors with which the nematode comes in contact, I wish to stress that it is easy to conclude that a cause-and-effect relationship exists between nematode behavior and a given soil property, especially when there is a striking correlation. But, as will be reiterated several times, soil factors are interrelated and as one factor changes, so does another. Because controlled monofactorial experiments relative to soil are rare, the investigator must use caution in relating results to cause and effect. This does not imply, however, that one factor cannot be responsible directly or indirectly for a certain nematode phenomenon.

Factors that are relatively easy to measure, or so it seems, are frequently employed in experiments. A good example is pH; many correlations of biological phenomena have been made with pH. But are the differences due to pH itself or to the factors related to it? In another instance, is the decline in the survival of a nematode due, at least at first, to a reduction in moisture per se or to the higher osmotic pressure that accompanies such a reduction? Knowledge of the environment in which an organism cannot live or thrive is often as important to understanding the limitations of population development as knowledge of those factors that promote optimum development. This is one of the bases in deciding the need for regulatory measures. We should not forget that differences that seem small to us may have definite effects on nematodes.

THE SOIL PROFILE

If a vertical section of soil is cut 3 to 5 feet deep, one will notice in most instances that the soil is divided into layers or horizons of different colors, textures, and thicknesses. This layering constitutes the soil profile and provides the series identity for the type. From top to bottom, the horizons are designated as the O, A, B, C, and R, or bedrock, horizons. Each layer may be subdivided. The O or organic horizon is at the surface and contains organic debris in various stages of decomposition but is still separated from the mineral portion of the soil. The A horizon is the topmost mineral portion, with the A-1 horizon frequently being dark due to the coating of mineral particles with the organic matter. The B horizon often has an accumulation of clays and iron and aluminum oxides. The C horizon is frequently dense and is generally outside the area of major biological activity, especially in most field crops. All horizons are not necessarily present, and transitions can occur between two adjacent horizons. The plow layer may or may not encompass the entire A horizon, and it may penetrate the B layer where the A horizon is shallow. The plow layer, of course, tends to become relatively homogeneous. The soil texture of the A horizon can be designated as a soil phase, such as Webster-clay-loam, which encompasses the series name as well as soil texture. A soil phase is based not only on the A horizon texture but also on the composition of underlying layers. These deeper layers greatly govern the drainage and leaching capacities of the profile. Although a soil may seem static and unchangeable over short periods, it is actually a dynamic entity and will change over time, thus gradually creating new habitats with resultant succession of organisms.

MINERAL COMPONENTS

The mineral fraction of the soil is composed of sand, silt, and clay. These are inherited from the parent material but may be changed with time by the action of two major groups of forces, physical and chemical. The former includes many weathering aspects such as abrasion by wind and water, often accompanied by thermal changes that cause cleavage of particles to varying degrees. Chemical decomposition is a result of one or more processes that include oxidation, reduction, hydrolysis, and carbonation. Microorganisms frequently are important in these chemical processes. The soils formed vary according to the nature of the parent material and the physicochemical forces that predominate at the time of soil formation, a lengthy process at best.

Soil texture and type often are used interchangeably, but they are not the same. Soil texture refers to the relative proportions of sand, silt, and clay in a given portion of soil. Type has lost its specific meaning and now refers in a general way to a group of soils with similar characteristics. In older literature, soil type referred to the lowest unit of soil classification that considered the several layers of the profile including color, organic matter, texture, drainage, and other properties. Unless qualified, we assume that the texture attributed to a soil is that of the A horizon. The soil separates are groups of soil particles based on diameter (Table 6.1). Several gradations of sand and sometimes clay are often delineated. Although other minerals may

Table 6.1. Soil separate sizes (U.S. Department of Agriculture system)

	Diameter	
	mm	Microns
Sand	2.00–0.05	2000–50
Silt	0.05–0.002	50–2
Clay	<0.002	<2

be present, sand and silt are dominated primarily by silicon dioxide and are relatively inert, especially in the temperate regions. Because of size differences, sand and silt differ in their moisture-holding capacities, bulk densities, and consequently the amount and size of pore space, which in turn affects the aeration and ease of drainage. Clay particles are the smallest of the soil separates, and many are colloidal. A few clays are amorphous, but most are laminated into plates with definite ionic arrangements. Naturally, different arrangements of lamination and chemical composition result in different properties. Thus there are different kinds of clays. Small clay particles provide an extremely large total surface area compared with those of silt and sand. The lamination of the particles provides an internal surface in addition to the external surface. This internal surface not only greatly increases the adsorptive capacity, but also is important as an increased source of ions, an important factor as we shall see when cation exchange is discussed.

We often refer to sand, silt, and clay fractions as separate entities, but they intergrate in nature. All fractions are important to nematode ecology, but they generally act in different ways. Unfortunately, when one presents the fractions as a mechanical analysis with no other information, the reader

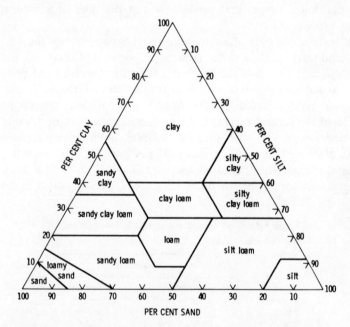

Figure 6.1. Diagram for determining soil texture, U.S. Department of Agriculture system.

is left with just that—a mechanical analysis of components that is devoid of structure and of the dynamic system from which the fractions originated. Also unfortunately, breakdown of the textural fractions provides little insight into the organic matter of the soil, which along with the clay is the most chemically and biologically active soil ingredient. We are becoming increasingly aware that chemically active portions of the soil are important in nematode ecology. The chemical aspects of soil in relation to nematode ecology will be treated in another chapter.

A scheme for placing boundaries on textures as implemented by the U.S. Department of Agriculture is presented in Figure 6.1. One can see that the percentage of the three components in any texture usually varies considerably. For example, clay-loam may have from 20 to 45% sand and still be called a clay-loam, a fact that may be important in describing nematode habitats. The proportion of sand will affect the moisture-holding capacity and aeration, drainage, and the microbial population, to say nothing of the growth of the host.

SOIL MOISTURE

Water not only governs the life processes of nematodes, it is also the important medium for active migration in soil and for passive dispersal by rivers, streams, irrigation canals, floodwaters, and in percolation. Nematodes are said to be essentially aquatic; most forms live in a film of water. Due to the width of many nematodes and to the pore sizes in most agricultural soils, this film is not a uniform one engulfing both the soil particles and nematodes. Except in soil containing little air, the nematode creates a bulge as the water film covering it continues around the soil particle if the diameter of the particles is larger than the nematode. More often, when soil particles are of smaller diameter than the nematode, nematodes move in a sheath of water drawn from the particles contacted, especially when the soil is near field capacity.

Soil moisture varies considerably as a result of many biolgical and physical actions. In the broad sense, some of the most obvious factors regulating soil moisture are the intensity and frequency of rainfall, the drainage pattern, and the precipitation/evaporation ratio. Other factors governing the amount of moisture in the nematode's environment include the amount of available soil moisture for the particular soil type, the stage of plant maturity, the length of the growing season, the water requirements of individual plants or crops, and elements relating to the individual characteristics of the leaves and roots.

It is well recognized that even though much attention is given to water per se, the changes of other properties such as gas tensions, osmotic pressure, temperature, and the biological community, all of which change with changing moisture conditions, may be as important as water changes in governing nematode activity. The addition of moisture usually reduces the available oxygen, stabilizes temperature, lowers osmotic pressure, and reduces salt concentrations in the soil. A depletion of moisture usually has opposite effects.

The retentive capacity of moisture in soils is so dependent upon the soil texture and other factors that determine the soil phase that they cannot be separated when studying soil water. The coarser the soil, the larger the pores; the smaller the total soil surface area, the less water retained. Clays, which are composed of extremely small particles, have an enormous surface area, small pores, and thus a great water-retaining ability. The amount or kind of life that soil water can support depends upon the soil composition and the ability of the organism to live at given moisture levels. Although plants vary in their ability to thrive at different moisture levels, a rule of thumb is that most cultivated plants do best at slightly below field capacity. Water is adequate and there is sufficient aeration for suitable gas exchange. Truly saturated soil is probably relatively rare as air is likely to be trapped within the soil. This could aid in survival of aerobic microorganisms for short durations but probably is inconsequential thereafter, as the oxygen is soon consumed and toxic substances that form may actually hasten the mortality of many microorganisms.

Since water in the soil is held by attractive forces between the water and soil particles (adhesion) or between the water molecules themselves (cohesion), the percentage of water in soil is not sufficient to indicate the energy relationships by which the water is held. The attractive forces (matrix for the soil and solute for the soil solution) create a tension or soil moisture suction. This tension can be expressed as centimeters of water, millibars, atmospheres, pF (the logarithm of soil moisture tension in cm water), or ergs per gram. These equivalents are depicted in Figure 6.2. Soil moisture curves can be constructed that depict the relationship between the soil moisture tension and the percentage of soil moisture for any one soil. Energy is involved in the attractive forces, and if the free energy of free water is zero, then the free energy of any bound water is less than zero, or negative. Water that is tightly bound has little free energy, or a great negative free energy. Surface relationships are important, and coarse soils, which have less total surface than finely textured soils, have less attraction for water. Thus, a coarse soil will hold less water at the same moisture tension than a finely textured one. The concentration of salts or other substances will also affect the free energy of water and consequently its capacity to evaporate. Tension at the soil particle–water interface of thin films is about 10,000 atmospheres, whereas that at the outer edge of the thickest films is about

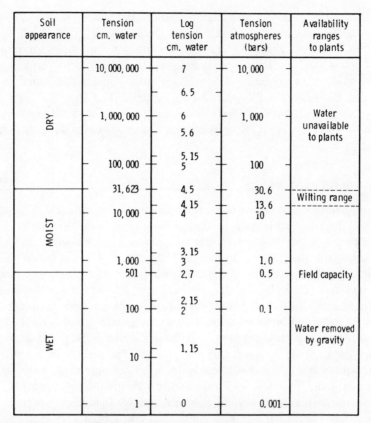

Soil appearance	Tension cm. water	Log tension cm. water	Tension atmospheres (bars)	Availability ranges to plants
DRY	10, 000, 000	7	10, 000	Water unavailable to plants
		6. 5		
	1, 000, 000	6	1, 000	
		5. 6		
	100, 000	5. 15 / 5	100	
MOIST	31, 623	4. 5	30. 6	Wilting range
	10, 000	4. 15 / 4	13. 6 / 10	
	1, 000	3. 15 / 3	1. 0	
	501	2. 7	0. 5	Field capacity
WET	100	2. 15 / 2	0. 1	Water removed by gravity
	10	1. 15		
	1	0	0. 001	

Figure 6.2. Soil moisture relationships.

1/3 atmosphere. Wet soils have low tensions, and at high tensions water is unavailable to organisms. In any event, nematodes live in pores formed among soil particles or soil crumbs. Consequently energy relationships of pores that nematodes cannot enter because of body size are of no importance.

The maintenance of soil moisture is a difficult problem, and many experiments doubtless have been in error in their design. This is predicated on the fact that, when water is added to a volume of soil below field capacity, the soil will be moistened to field capacity in an expanding area until the free water added is completely used. The soil beyond this expanding field capacity area will remain at the original soil moisture level below field capacity for a very long time. Thus, it is impossible to add water to the top of a volume of soil and expect the whole volume to come to the equilibrium at a point below field capacity. Sometimes investigators maintain different ranges of moisture levels by bringing a soil to field capacity and allowing it to dry to various moisture levels below field capacity. This cycle then is repeated throughout the experiment. In attempts to maintain moisture levels below field capacity, moisture sometimes is added to various depths by tubes. This, of course, will give many little pockets moistened to field capacity. Sometimes soil is spread out, air dried, and sprinkled with moisture to theoretically give different moisture levels below field capacity, but the same problems continue to exist, and such a system is difficult to maintain.

SOIL TEMPERATURE

Soil temperature and moisture are the most studied parameters in nematode ecology. This is due largely to the seeming ease of measurement, but in actuality the problems involved can be complex. It is easy to insert a thermometer into the soil or to use soil thermographs, but since thermal gradients exist both vertically and horizontally, giving rise to microclimates, experimenters must be careful not to rely too much on single point measurements. Sufficient soil temperature readings have been taken by meteorologists and other scientists that many generalities can be made, but generalities are often misleading. Air temperatures as recorded in weather bureau data should be used only for general climatic trends because soil temperatures can differ considerably from air temperatures. For example, snow is a great insulator, and unfrozen soil covered deeply with snow can remain unfrozen even though air temperatures are well below freezing.

As with other parameters, soil temperature affects activities of organisms, and it greatly infuences or regulates, and is regulated by, other parameters. Different organisms are affected differently; whereas a certain temperature itself may limit activity, in conjunction with other parameters it may have a modifying or accentuating effect. Temperature is important in the formation of soil and the life therein. Freezing and thawing affect soil structure. Since water has one of the largest heat-holding capacities of any substance, wet soil warms up more slowly than less wet or dry soil.

Nearly all heat in the soil comes ultimately by radiation from the sun although a minor portion comes by conduction from areas deep within the earth. Within the earth's atmosphere and soil surface layers, however, many factors govern the amount of heat available to the soil. These include the angle of incidence of radiation, topography, the amount of fog, dust, clouds, convection currents, air drainage evaporation, rainfall, transpiration from plants, and the plant cover. Internally, the biological activity, moisture, texture, structure, salt concentration, and radiation from the soil affect soil temperature directly or indirectly.

Soil temperature is one of the most constantly changing factors that a nematode encounters. Diurnal fluctuations vary greatly depending upon the soil type, texture, moisture, atmospheric conditions, latitude, elevation, and season of the year. Fluctuations usually are greatest at the soil surface and become less pronounced in deeper layers until at a depth of about 3 feet there is little diurnal fluctuation. Seasonal fluctuations may occur much deeper in the temperature zones than in the arctic or tropical areas. Also, the drier the soil, the greater the fluctuations. Soil surface temperature and air temperature are approximately the same, but there is a lag in the maximum and minimum temperatures in deeper layers.

AERATION

As a rule of thumb, about one-half of a mineral soil consists of solid particles and the remaining portion contains liquids and gases. The gaseous portion of the soil also varies widely as water fluctuates widely in most agricultural soils. Thus it is difficult to speak of any normal amount of gas. The composition of the gaseous phase also varies within and between habitats, depending upon such factors as rainfall, drainage, temperature, vegetational requirements, and microbial activities. In general, oxygen is somewhat lower in soil than in atmospheric air, carbon dioxide is much higher, and nitrogen is about the same. The carbon dioxide content of atmosphere averages about 0.003%, but that of soil air is usually between 0.1 and 1.0% near the surface, and it can be higher, to as much as 20%, in the deeper layers. The percentage of carbon dioxide in soil air is usually greater, or the oxygen is usually lower, in summer than in winter, in vegetative than in nonvegetative soil, in fertilized than in nonfertilized soil, in subsoil than in topsoil, in fine than in coarse textured soil, and in wet than in dry soil. Gas exchange occurs by diffusion and mass flow. Diffusion is continuous, but mass flow is sporadic.

ORGANIC MATTER

The soil organic matter is important in water retention, cation exchange capacity, soil aggregation, erosion control, plant nutrition, and as a substrate for microorganisms. There is increasing evidence that the organic matter is important in nematode ecology, and therefore nematologists require a basic knowledge of the subject.

The organic matter of the soil consists of all living or once living matter, large or small, simple or complex. Large living and dead portions such as roots, earthworms, and some insects are removed before analyses. Smaller life such as microorganisms are included in the analyses largely because of separation difficulties. The residues-form of organic matter includes dead parts of plants and animals and excrement. These materials are subject to rather rapid decomposition, especially if the residues are near the surface where there is good aeration and if moisture and temperature are favorable. Another form of organic matter is humus, which results from slow and incomplete decomposition. Humus has more consistent chemical and physical properties than the organic residues. The kind, rate of formation, and persistance of humus depends largely on the source and composition of organic materials, temperature, aeration, mineral content of the soil, and microorganisms, with the latter playing an important role. The exact chemical nature of humus is not known; it is a mixture of compounds that varies somewhat in composition. Nearly half consists of ligninlike compounds, about one-third of amino acids, and the remaining portion is mostly carbohydrates plus some cellulose, hemicellulose, fat, wax, and miscellaneous compounds. Humus is often classified as raw, nutrient, and true or neutral humus, progressing from the least to the more advanced stages of decomposition. Many of the true humus particles become coated with clay elements, forming clay-organic complexes. Organic particles are charged and are important in the cation exchange. The activities of soil organism produce carbon dioxide, a large amount of which diffuses into the atmosphere. Some carbon dioxide in the soil, however, is used in the formation of carbonic acid, carbonates, and bicarbonates.

CATION EXCHANGE

The cation exchange capacity (C.E.C.) is the total number of cations that can be adsorbed on a soil and are exchangeable with cations in the soil solution. Thus, there is little or no degradation of the soil solids. This capacity is determined at neutrality (pH 7) and is expressed as milliequivalents per 100 grams of soil. Quantitatively the most important exchangeable cations in the soil are usually calcium, magnesium, potassium, sodium, aluminum, iron, manganese, and hydrogen, although occasionally other cations assume importance. Hydrogen, aluminum, iron, and manganese are the main contributors to soil acidity, while the others make a soil more alkaline or less acid. Clay and organic matter are the two more chemically active parts of the soil, and the cation exchange capacity is largely a reflection of these two entities. Since sand and silt are largely inert, these fractions contribute little to the cation exchange capacity. Thus soils consisting mainly of sand and silt will have a low cation exchange capacity if there is little organic matter present. Since different clays have different capacities to adsorb ions, one cannot judge the C.E.C. by knowing the percentage of clay in a soil. Due to its structure and comparatively large particle size, kaolinite has a much smaller C.E.C. than montmorillonite, whose particles are much smaller and have a looser internal structure, which greatly increases the adsorption capacity. But such a configuration is not representative of all clays. Although C.E.C. may seem a constant characteristic for many· soils, it can vary from time to time depending on the nature of the colloidal material available and the pH of the soil. The most variable of this material is the colloidal organic matter, which will depend upon the nature and rate of decomposition.

pH

The effect of soil pH on the occurrence and biology of nematodes has received little attention. There are some contradictions, indicating complicating factors, which are bound to exist in such a complex environment as soil. Studies of pH and nematodes usually are based on one of two approaches. One is to collect a large number of soil samples covering a wide pH range and then to attempt quantitative and qualitative correlations with nematodes. The other is to adjust the pH of a given soil and then follow the nematode behavioral changes. In the field, sulfur usually is added to make the soil more acid, and lime to make it less acid. In the laboratory, H_2SO_4 and $Ca(OH)_2$, respectively, are commonly used.

The pH is expressed as the negative logarithm of the hydrogen ion concentration, although it actually expresses the activity of hydrogen ions. Thus

$$pH = -\log [H^+] = \log \frac{1}{[H^+]}$$

so that $H^+ = 10^{-pH}$. A hydrogen ion concentration of 10^{-8} is expressed as pH 8. The greater the hydrogen ion concentration, the lower the pH, and consequently the lower the hydroxyl $[OH^-]$ concentration, since the two are in equilibrium. pH of the soil solution is not the simple and straightforward phenomenon that one might assume from reading a figure on a pH meter. The factors responsible for pH are complex and depend greatly on the soil solution and the type of colloids present, the adsorbed bases, and the microbiological activity, which in turn is influenced by pH. The pH also affects the availability of minerals and the physical properties of the soil. The hydrogen ions may come from hydrogen directly or from the hydrolysis of aluminum compounds.

Regionally the soil reaction is quite variable. The pH of the soil solution in humid regions ranges from about 5.0 to 7.5 and in arid regions from approximately 6.5 to 9.0. The pH may extend beyond these extremes in many instances. Soil tends to be acid in humid regions because there is enough precipitation to leach out appreciable amounts of exchangeable bases (cations other than hydrogen and aluminum). In arid regions there is not enough rainfall to leach the exchangeable bases, which results in a high base saturation and a preponderance of hydroxyl ions. Thus hydrogen and aluminum ions tend to dominate acid soil. The pH of the soil is not constant, although soils that have high exchange capacities such as those with considerable amounts of clay or organic matter are generally well buffered. The pH can vary considerably within a field. This may be due to the heterogeneity of the soil itself, to the uneven distribution of organic residues, or to the microbial activity through the production of carbonic, nitric, and other acids. From a microbiological standpoint, it is well to remember that the hydrogen ions are adsorbed on soil particles more strongly than are hydroxyl ions; therefore, the former may be more concentrated at the colloidal interfaces than in the outer portions of the water film. Different readings are obtained when different techniques are used for measuring pH. It is important to state the method used so that proper comparisons and interpretations can be made. Readers who wish to become better acquainted with the intricacies of pH and their effects on soils and microbes are encouraged to consult texts on the subject.

OXIDATION-REDUCTION POTENTIAL

The oxidation-reduction potential, or redox potential as it is known, is sometimes useful in soil studies although it has received only superficial attention relative to nematodes. Basically the oxidation-reduction potential is reflected in the soil solution and for the most part is the result of electron transfer through the action of metabolic processes of soil micro- and macroorganisms. The presence of several redox systems contributes to the difficulties in its measurement and in interpretation of results, and the researcher should be well acquainted with these before attempting experiments and interpreting results relative to nematodes.

PHYSICAL AND CHEMICAL PARAMETERS

Environmental factors never operate singly, but in
conjunction with some or all of the others.

So far we have briefly discussed nematodes in a diversity of habitats and
have said a little about population change. Although the host is of primary
importance in a nematode's becoming established, once a susceptible host is
present other factors largely control the population size of the parasite.
Some interrelationships of nematodes with the environment are depicted in
Figure 7.1. We will turn now to those environmental resistance factors that
limit the development of nematode populations. In doing so, we should
never forget that these parameters, even though they are discussed singly,
almost always are in a complex relationship with other factors acting
simultaneously. The effect of temperature is influenced often by moisture.
Predators and parasites interact with many physical, and chemical, and biological properties. Thus one parameter's effect on nematodes may vary with
the environment.

TEXTURE AND MECHANICAL ANALYSIS ————————————————

Some nematodes occur, increase to large populations, or are involved in
plant unthriftiness more in soils of certain textures than in others. Information on the soils in which nematodes have been found to occur is given in
Table 7.1. Although some reports seem to be based on little more than
casual observations, the association must have made a suitable impression

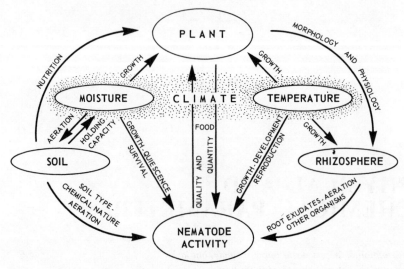

Figure 7.1. Interrelationships among plant-parasitic nematodes, plant, climate, and soil environment. (From National Academy of Science, Control of plant-parasitic nematodes. 1968. Courtesy of S. D. Van Gundy.)

upon the investigator to be worthy of record. When enough observations and experiments agree, a realistic and general phenomenon is probably indicated. Frequently, however, apparent contradictions occur, but since other factors can influence population size this is to be expected. Inherent factors such as pore size, total pore space, crumb structure, and percentage of organic matter, and external factors such as the amount of water received, cropping practices, frequency of cultivation, fertilization practices, host susceptibility, vigor of the host, pesticide applications, and the activity of associated organisms can influence nematode behavior. Therefore, in efforts to determine generalities relative to soil texture, many samples representing a variety of soils should be analyzed. A conclusion based upon three or four samples or fields should not be treated as depicting a generality, because nematodes are not uniformly distributed within or between fields of the same soil texture or type. Table 7.1 should be examined with this in mind, along with the fact that the authors usually made no claim that their observations reflected a generality.

Soil particles frequently are studied separately in experimental work, but most soils are heterogeneous in nature. Interpretations must be made cautiously when laboratory and field results are being compared. Evidence is accumulating that the percentage of a given particle size per se is not as important as the interrelations among separates or other edaphic factors.

As examples, Kimpinski and Welch (1971) found that nematode numbers in three Manitoba soils were correlated with nitrogen, potassium, and soluble salts, but not with the percentage of sand, silt, or clay. Few correlations of nematode numbers occurred with percentages of sand, silt, and clay in soybean fields, but there were many correlations with other soil factors, including texture (Norton et al., 1971). Cohn (1969), however, correlated the volume of five *Xiphinema* species with soils of three different textures and found that the nematode with the smallest volume (*X. mediterraneum*) attained its densest population in clayey soils. The next densest populations were in loamy soils, but few nematodes occurred in sandy soils. The opposite was generally true for the larger nematodes. It is not known, however, how much the populations were governed by particle size, crumb size, moisture, aeration, or other factors.

In spite of the volume of work on particle size and texture, there is evidence that soil structure is more important for some nematode increase. Jones et al. (1969) and Jones and Thomasson (1976) contend that nematodes usually occupy only macropores, those greater than 30 μm in diameter, and that micropores, or those less than 30 μm in diameter, are generally not accessible to nematodes. They state that the former are usually between aggregates or crumbs, while micropores usually occur within aggregates or crumbs. As partial support for their argument, Jones et al. (1969) observed that *Xiphinema diversicaudatum*, a large species, occurred in the macropores between impenetrable aggregates. *Trichodorus* and *Longidorus* spp. occurred significantly only in soils with more than 80% sand.

The most extensive work on the physics of soil particles and moisture relative to nematodes has been that of Wallace (1958a,b,c; 1971a) and his colleagues. The topic will be mentioned only briefly here. The size of particles or crumbs affects nematode movement directly in that it controls the pore size. Small pores prevent nematode passage; if the pores are too large, leverage becomes a problem. Wallace (1958a), experimenting with 11 nematode species, concluded that the maximum mobility is obtained when the nematode length is three times the diameter of the soil particles. This is assuming that moisture is optimum since different suctions will also affect the mobility of nematodes. If the film thickness in which the nematode lives is too great, there is slippage; if the films are too thin, there is too much friction and mobility is inhibited. Thus there is probably an optimum soil particle size and an optimum film thickness for each nematode species. Walace (1958a) believes that this thickness is 2 to 5 μm for *Heterodera schachtii*. Ponchillia (1972) found that migration of *Xiphinema americanum* was significantly greater in particle size fractions of 150 to 250 μm or 500 to 700 μm than in soils having particle sizes of 75 to 150 μm or 700 to 1000

Table 7.1. Soil textures and nematode occurrences

Nematode	Soil condition	Reference
Aglenchus costatus	Wet heavy soils	Geraert, 1967
Aphelenchus avenae	No preference for soil type	Geraert, 1967
Basiria graminophila	Found in heavy soil	Thorne and Malek, 1968
Belonolaimus euthychilus	Mostly in sand, swamp, and alluvial soils	Rau, 1963
B. gracilis	Sand (probably included *B. longicaudatus*)	Holdeman, 1955
B. longicaudatus	Sand, loamy-sand	Miller, 1972
B. longicaudatus	Increase occurred only with minimum of 80% sand and a maximum of 10% clay	Robbins and Barker, 1974
B. nortoni	Sand	Rau, 1963
Boleodorus thylactus	Wet heavier soils	Geraert, 1967
Criconemoides caelatus	Sand	Raski and Golden, 1965
C. quasidemani	Found in sandy soil	Raski and Golden, 1965
C. xenoplax	Increase greater in sandy soil	Seshadri, 1964
Ditylenchus dipsaci	Movement optimum in soil 250 to 500 μm	Wallace, 1958b
D. dipsaci	Clay	Seinhorst, 1956
D. dipsaci	Dry sandy soils	Geraert, 1967
Dolichodorus silvestris	Found in dry, shallow, shaly soil	Gillespie and Adams, 1962
Dorylaimids	Least in heavy soils, most in light soils	Norton et al., 1971
Helicotylenchus digonicus	Mainly in heavier soils	Brzeski, 1970a
H. dihystera	Populations higher in clay-loam than sandy-loam	McGlohon et al., 1961
H. pseudorobustus	Negatively correlated with silt; positively correlated with clay	Norton et al., 1971

Species	Soil preference	Reference
H. pseudorobustus	No preference for soil type	Geraert, 1967
Hemicriconemoides chitwoodi	Fine sandy-loam best compared with silty-clay-loam and loamy-sand	Chang and Raski, 1972
Hemicycliophora arenaria	Greatest reproduction in sandy soil	Van Gundy and Rackham, 1961
Heterodera avenae	Silty-clay in low areas	Putnam and Chapman, 1935
H. avenae	Sandy soils	Southey, 1956; Geraert, 1967
H. avenae	Multiplied best in sandy-loam compared with heavier textured soils	Meagher, 1968
H. punctata	Found in black sandy-loam soils	Russell, 1927
H. punctata	Wet heavy soils	Geraert, 1967
H. schachtii	Densities greatest in clay-loam compared with nine other soils	Caveness, 1958
H. trifolii	Wet heavy soils	Geraert, 1967
H. trifolii	A-loamy-clay-sand better than sand or a sandy-clay-loam	Norton, 1967
Longidorus breviannulatus	Mostly in soil with more than 90% sand	Norton and Hoffmann, 1975
L. brevicaudatus	Populations greater in sandy and loam soils; was not found in clayey soils	Cohn, 1969
Meloidogyne incognita acrita	Decreased yields in loamy-sand and sandy-loam compared with silt-loam	O'Bannon and Reynolds, 1961
M. javanica	Infestations light in finely textured soils but increased with coarser soils	Sleeth and Reynolds, 1955
M. javanica	Soils of 61–75% sand with a silt to clay ratio of 1:1, 2:1, or 3:1 are best for infectivity	Elmiligy, 1968
Meloidogyne spp.	Disease more prevalent in sandy-loam than in heavier soils	Sasser, 1954
Merlinius brevidens	No preference for soil type	Geraert, 1967
Nacobbus aberrans	Mostly in sandy soils	Thorne and Schuster, 1956
Paratylenchus projectus	Populations greatest in clay-loam than in sandy-loam or fine-sandy-loam	McGlohon et al., 1961

121

Table 7.1. *(Continued)*

Nematode	Soil condition	Reference
P. projectus	Mainly in heavier soils	Brzeski, 1970a
Pratylenchus brachyurus	Best in sandy-loam → loam → sand	Endo, 1959
P. coffeae	Largest populations from roots growing in relatively heavy soils	Riggs et al., 1956
P. crenatus	Large populations in sandy soil	Dickerson et al., 1964; Geraert, 1967
P. hexincisus	More abundant in heavier soils	Norton, 1959; Thorne and Malek, 1968
P. neglectus	Mainly in heavier soils	Brzeski, 1970a
P. penetrans	Soil not a factor	Dickerson et al., 1964
P. penetrans	More in light sandy soils	Mai and Parker, 1956; Mountain and Boyce, 1958; Parker and Mai, 1956; Slootweg, 1956; Springer, 1964
P. scribneri	More in sandy-loam than silty-clay-loam	Thomason and O'Melia, 1962
P. thornei	Utah infestations only on heavy clay-loam soils	Thorne, 1961
P. thornei	Maximum numbers in clay (Australia)	Grandison and Wallace, 1974
P. vulnus	Heavy clay	Knobloch, 1975
P. zeae	Disease worse on light sandy soils	Sher and Bell, 1965
P. zeae	More in light sandy soils	Jenkins et al., 1957
	Migration greatest in sandy-loam → loam → clay	Endo, 1959
Rotylenchus fallorobustus	No preference for soil type	Geraert, 1967
R. goodeyi	No preference for soil type	Geraert, 1967
R. robustus	Dry sandy soil	Geraert, 1967

Species	Soil preference	Reference
Trichodorus christiei	Movement less in clay-loam than in sandy-loam	McGlohon et al., 1961
T. christiei	Populations greatest in sandy-loam → loam → silty-clay-loam	Thomason, 1959
T. pachydermus	Common in sandy soils	Sol and Seinhorst, 1961
Tylenchorhynchus dubius	Common in lighter soils	Brzeski, 1970a; Geraert, 1967; Sharma, 1968
T. lamelliferus	Wet heavy soils	Geraert, 1967
T. maximus	No preference for soil type	Geraert, 1967
T. microdorus	Dry sandy soil	Geraert, 1967
Tylenchulus semipenetrans	10–15% clay	Van Gundy et al., 1964
T. semipenetrans	Best development of juveniles into adults in loamy-sand	Baines, 1974
Tylenchus agricola	No preference for soil type	Geraert, 1967
T. baloghi	No preference for soil type	Geraert, 1967
T. leptosoma	Dry sandy soil	Geraert, 1967
T. magnidens	No preference for soil type	Geraert, 1967
T. minutus	Mainly in heavier soils	Brzeski, 1970a
Xiphinema americanum	Negatively correlated with clay	Norton et al., 1971
X. bakeri	Widespread in sandy soils of raspberry in British Columbia	McElroy, 1970
X. brevicolle	Populations greatest in sandy soils	Cohn, 1969
X. diversicaudatum	Populations greatest in sandy soils	Cohn, 1969
X. diversicaudatum	Occurred in medium to heavily textured soils; not found in sandy soils	D'Herde and van Den Brande, 1964
X. index, X. brevicolle, Longidorus africanus	Higher populations in heavy (32% clay) than in light (8% clay) soil	Cohn and Mordechai, 1970
X. italiae	Populations greatest in sandy soils	Cohn, 1969
X. mediterraneum	Populations greatest in clayey → loamy → sandy soils	Cohn, 1969

μm. He also found that migration of the nematode was greater in silt-loam and loamy-sand than in silty-clay or clay-loam. It should be remembered that in such experiments the soil was disturbed initially, and thus the structure is destroyed and compaction may not be the same as in the field.

Decrease in nematode numbers with increasing soil depth is often associated with a decrease in the number of kinds of roots. Although roots are important factors in vertical distribution patterns, soil particle size and the resulting changes in pore space may be contributing. Soil at the lower depths is compacted more than that in the plow zone. Besides smaller pores, drainage and aeration are often poorer in deeper zones than in surface zones.

MOISTURE

Regional Occurrence

Moisture and temperature are often coinfluential, and it is usually difficult to separate the effects of the two. Some nematodes that are indigenous to the arid regions of the western United States have not been recorded from the humid areas of the eastern states. Unfortunately we do not have the necessary information to separate the limiting factors. Two species of nematodes about which we can speculate reasonably well as being restricted by moisture spend much of their lives above ground, namely *Ditylenchus dipsaci* and *Anguina tritici*. In the United States, *D. dipsaci* has been reported most frequently along the more humid areas of the eastern and western seaboards and the irrigated regions of the West. It is occasionally serious in the Midwest, however. Although the preadult stage is resistant to desiccation, evidence indicates that the nematode is governed greatly by moisture. Seinhorst (1956) states that all Dutch rivers and marine clay soils seem to be infested with the stem nematode, and that the disease on onion is persistent in heavy clay soils but not in sandy ones. Wallace (1962) reasoned that heavier concentrations of *D. dipsaci* would allow for greater lateral transport, and that in the absence of rainfall the surface populations would decline. He found that after a rain there was a marked increase in the number of *D. dipsaci* at the surface. While testing two populations of the nematode on alfalfa at three different temperatures, Barker and Sasser (1959) found that infection and reproduction increased at high soil moistures. As temperature and moisture are so interrelated, the effects of the former cannot be eliminated. The nematode reproduced better on alfalfa at an air temperature of 18 C than at 13 C or 24 C. Griffin (1968) found that although maximum infection was at 20 C it could occur from 5 to 30 C.

The even more restricted geographical range of *Anguina tritici* in the United States certainly cannot be attributed to the distribution of the host. The nematode has been found almost entirely in the more humid regions of the southeastern United States. As mentioned in Chapter 1, the persistence of the nematode outside of the Southeast is questionable. Although historical factors, possible varietal isolation, and the narrow host range must be considered, moisture seems the most plausible reason for the restricted distribution of this nematode. Collis-George and Blake (1959) concluded that maximum expulsion of *A. agrostis* from galls and movement of juveniles occurred near field capacity. If this is true for *A. tritici,* the optimum soil moisture would not prevail as long near the soil surface in the drier midwestern and western states as in soils of the eastern states.

It is well known that some nematode stages, such as the preadult of some *Anguina* species and the fourth or "wool" stage of *Ditylenchus dipsaci,* are more resistant than others to desiccation. Less dramatic are the resistant stages of *Paratylenchus* that are tolerant to desiccation (Reuver, 1959; Rhoades and Linford, 1961b). However, Rössner (1971) and Simons (1973) have shown that nematode survival in dry soil is greater than is usually acknowledged. They found that wetting of the soil before sampling often revives many nematodes that would not otherwise be recovered.

Seasonal Changes

The addition of moisture, either by rainfall or irrigation, usually has a profound influence on the behavior of nematodes. The nature of these changes depends, of course, on the amount of moisture added and on the edaphic and climatic factors existing at the time. In a dry season, rainfall usually results in increased egg hatch and activity following quiescent periods during drought. Seasonal dry spells occurring during the nematode's active period may cause sharp dips in the population, but some nematodes have a remarkable ability to recover, even those that are sensitive to desiccation, as found for *Xiphinema americanum* (Norton, 1963). Griffin and Darling (1964) also found the fewest *X. americanum* when high temperatures and minimum moisture prevailed, but the nematodes increased with increased moisture and cooler temperatures. Perhaps some of these investigations should be reexamined in view of the results of Rössner (1971) and Simons (1973) on nematode survival in dry soil. While studying nematode populations under adequate moisture conditions in Alabama, Minton et al. (1960) found that, with the possible exception of *Hoplolaimus tylenchiformis,* rainfall had little effect on population changes of plant-parasitic nematodes in cotton. Van Gundy (1958) speculated that the interval between irrigations of citrus infected with *Tylenchulus semipenetrans* is important in the life cycle of the nematode because free water is necessary to flush the juveniles from the egg masses.

When moisture becomes excessive, owing to either rainfall or irrigation, nematode numbers often decline. For example, Hollis and Fielding (1958) found an inverse correlation of *Tylenchorhynchus* spp. with the amount of flooding and rainfall in Louisiana. Increased frequency and amount of irrigation around citrus also decreased numbers of *Hemicycliophora arenaria*, and this decrease was linked to oxygen diffusion rates (Van Gundy et al., 1968). Seven days was required between irrigations to allow sufficient oxygen to diffuse throughout the top 24 inches of soil. Three-day irrigation intervals did not allow enough oxygen diffusion for nematode reproduction or root growth (Figure 7.2). Cralley (1957) reported that rice seeded in water became less severely infested with *Aphelenchoides besseyi* than when drilled in and flooded later. This control possibly is due not only to suffoca-

POPULATIONS OF Hemicycliophora arenaria RELATIVE TO TEMPERATURE AND MOISTURE

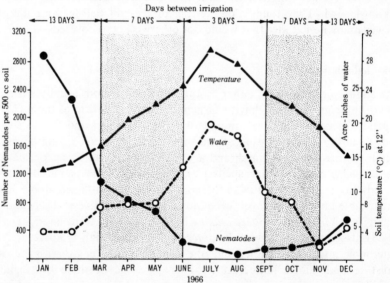

Figure 7.2. Numbers of *Hemicycliophora arenaria* in the top 30 in. of soil; soil temperature at 12 in. and acre-inches of water per month during 1966. (From Van Grundy, S. D., F. D. McElroy, A. F. Cooper, and L. H. Stolzy. 1968. Influence of soil temperature, irrigation and aeration on *Hemicycliophora arenaria*. Soil Sci. 106:270–274. Reprinted with permission of the author and The Williams & Wilkins Co., Baltimore.)

tion but also to the plants escaping the disease as the nematode densities become greatly diluted by flushing before invasion.

Small moisture differences are also important. Corn roots were penetrated by *Pratylenchus penetrans* and *P. minyus* at moisture tensions of 0 to 1000 cm of water, but the optima were between 10 and 100 cm of water. This means that more invasion occurs on the wet side of field capacity, which is roughly 500 cm of water tension, than on the dry side, as illustrated in Figure 7.3 (Townshend, 1972). On the other hand, it might be expected that at least some nematode species that thrive in upland soils would do well at somewhat below field capacity. Moisture often regulates the amount of aeration. Van Gundy et al. (1962) found that *Xiphinema americanum* survived the most poorly at low oxygen levels. In the same pattern, the nematode was found to occur in the greatest numbers in Iowa prairies in the well-drained slopes (Schmitt and Norton, 1972). A similar situation was found with *Rotylenchulus reniformis* in that infectivity was greatest when soil moisture was maintained at just below field capacity (Rebois, 1973b). (The term field capacity, although commonly used, is inaccurate in most laboratory work because the water column is broken during soil preparation.)

In the studies by Townshend (1972), low bulk densities favored penetration by the nematodes compared with high bulk densities. Since bulk density affects aeration, the amount of available water, and the resulting stress on the plant, the author correctly points out the possible differences in symptom expression of plants in different soil textures, even though similar nematode populations are present. Similarly, Santos (1973) found on interaction among temperature, moisture suction, and time on the mobility of *Meloidogyne* spp., which indicates that the optimum for any one factor is influenced by the others.

Moisture and Nematode Establishment

As far as is known, water is always a prerequisite for hatch, but other factors must be favorable. Dropkin et al. (1958) found a marked inhibition of hatching of *Heterodera rostochiensis* juveniles in solutions of 0.2 to 0.3 M NaCl, 0.3 M KCl, and 0.2 M $CaCl_2$, and complete inhibition of hatching at higher concentrations of these salts. Hatch inhibition, however, was overcome by subsequently placing the cysts into water. An optimum soil suction exists for the emergence of *Heterodera schachtii* juveniles from cysts. Emergence is low when a low suction occurs in the soil, but as suction increases so does the emergence, up to very dry conditions where emergence again decreases (Wallace, 1959a). A dry period sometimes results in

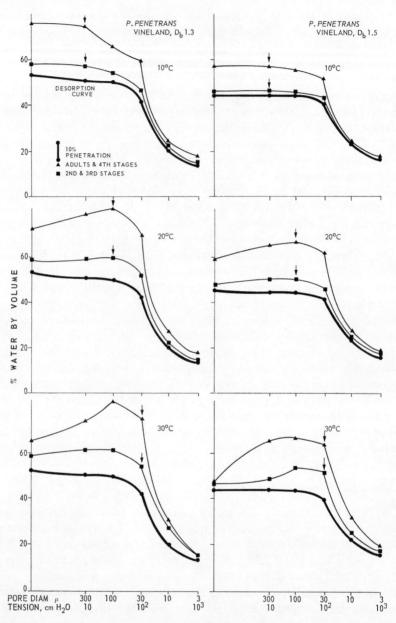

Figure 7.3. Percentage penetration of corn roots by *Pratylenchus penetrans* in Vineland silt loam at different bulk densities, moisture tensions, and temperatures. The arrows indicate the point of peak penetration. (From Townshend, *Nematologica*, 1972, p. 206.)

increased hatch when eggs are subjected to water, possibly by promoting maturity. A high percentage of *Heterodera glycines* brown cysts and more emergence from brown cysts were correlated with moisture stress on soybean plants (Hamblen and Slack, 1959). Fewer brown cysts and less juvenile emergence occurred if adequate or excessive soil moisture existed. This is consistent with results of Wallace (1959a), who found that, irrespective of the individual particle size of three grades of sand, maximum emergence of *H. schachtii* occurred when most pores were empty of water. These findings indicate the importance of knowing the moisture distribution in the pores. This was also emphasized by Fidler and Bevan (1963), who found that reproduction of *H. avenae* was limited by availability of pore space.

With the more exposed nature of root-knot juveniles, one might expect that desiccation is important in their life cycle. Peacock (1957) reported that root-knot juveniles and egg masses were susceptible to desiccation but that the egg masses were also intolerant of excessive moisture. The critical moisture content of the sterile loam soil used was 3 to 4% below that in which juveniles and eggs were unable to survive, except when protected by plant tissue. Juvenile survival was reduced slightly and egg survival was reduced drastically at the moisture holding capacity of the soil. These laboratory observations were confirmed by field work, which indicated that simple cultivation could reduce soil moisture and exert a practical measure of control. Viability of encysted juveniles of *Heterodera rostochiensis* free from soil and stored at 24 or 30 C declined more rapidly at 88% and 0 to 1.5% relative humidities than at other humidities tested, but generally viability was not affected by relative humidity at 4.4 C (Lewis and Mai, 1960). In contrast, unprotected juveniles did not survive exposure for 243 days at relative humidities of 3, 20, 88, or 100% at either 24 or 4.4 C. Wallace (1962) found that *Ditylenchus dipsaci* could not survive drying in 50% relative humidity equivalent to a tension of pF 6 for 34 days.

Because of the difficulty of maintaining constant soil moisture, little direct evidence can be gained on moisture and nematode increase. Kable and Mai (1968a) found that the population increase of *Pratylenchus penetrans* was greatest at moderate soil moisture tensions (pF 2 to 3) but that there was an interaction with soil moisture and soil type; the greater the amount of silt and clay present, the greater the soil moisture tension necessary for plant growth.

TEMPERATURE

With the development of constant-temperature tanks, great importance was placed on cardinal temperatures in discussions of plant disease development. Many excellent contributions were made, but at times more was read into cardinal temperatures than was actually there. The same has been true to some extent in plant nematology. Although there is value in knowing the particular temperatures at which an experiment was conducted, there are drawbacks in relying on them too much. A cardinal temperature might become impressed in the reader's mind, but in fact satisfactory reproduction may occur over a wide range of temperatures that would give a nearly flat peak on a temperature curve. Also, there may be two maximum peaks, as found by Thomason and O'Melia (1962) for *Pratylenchus scribneri.* Cardinal temperatures may reflect the overall result but do not bring out the fact that different stages in the life cycle of a nematode and/or disease development have different temperature requirements, as with *Meloidogyne ovalis* (Table 7.2). Such experiments are usually conducted under highly artificial and controlled conditions, which may partly explain the differences in the results obtained by different authors using the same nematode species. Different physiological races also may account for some discrepancies. The fact that the same organism may behave differently in different environments is illustrated with *Pratylenchus neglectus*, which reproduced best at different temperatures depending upon whether or not *Verticillium dahliae* was present. Thus such information should be used cautiously. Table 7.2 should be examined not so much for the effects of individual temperatures on nematodes as for general patterns.

It will be noted that most nematodes do best at temperatures between 25 and 30 C, but there are exceptions. A notable one is *Ditylenchus dipsaci*, which is generally favored between 15 and 18 C. This may account for the general severity of infections it causes in the spring and fall. The information in Table 7.2 is interesting and valuable. It should be realized, however, that temperatures may play different roles or are more or less effective in some environments than in others. When cardinal temperatures are coupled with survival temperatures, a clearer picture may unfold regarding geographical distribution and population patterns.

As might be expected, the effects of temperature become evident early in the life cycle of the nematode. Bird (1974) reported that exposing eggs of *Meloidogyne javanica* to 46 C for 10 minutes suppressed embryogenesis and hatching. Such phenomena may have survival value during periods of moisture stress.

It was mentioned in Chapter 2 that some nematode species have a temperature preferendum for migration and that migration of *Ditylenchus dipsaci* was optimum at the previous storage temperature (Croll, 1967). Griffin (1974) found similar results for invasion of the nematode into alfalfa. Nematodes were acclimatized at four temperatures ranging from 15 to 25 C for 6 months and then inoculated onto seeds and maintained at four temperatures ranging from 15 to 30 C. Although invasion occurred at all temperatures tested, it was greatest at or near the acclimation temperature. A greater percentage of plants were invaded at 15 or 20 C than at 25 or 30 C. Such phenomena might account for some of the variability reported by various workers for a given species of nematode, although it is probably not the only factor involved. But the works of Croll and Griffin illustrate the internal variability of nematodes in the disease process.

The use of accumulated temperatures relative to nematode development has value in predicting the number òf potential propagules, the length of the life cycle at different temperatures, and thus the number of life cycles in a season, other factors being favorable. Tyler (1933a) found that development of *Meloidogyne* sp. to egg laying required from 6500 to 8000 heat units, where a heat unit was a degree Centigrade above 10 C acting for 1 hour. Milne and DuPlessis (1964), using heat units as degrees Centigrade above 7.5 C over 1 hour, found that 9300 C-hours was necessary for *Meloidogyne javanica* to complete its development from second-stage juvenile to egg in the field. The onset of egg production of *M. hapla* on lettuce in the field was predicted within ±4 days by calculating heat units from a threshold temperature (Starr and Mai, 1976b). Jones (1975) recorded that accumulated average day degree temperatures above a basal development temperature plotted over time resulted in curves similar to the development of *H. schachtii*. The basal or threshold temperature below which nematode development does not occur or is negligible in most instances would, of course, vary with the species. Accumulated temperatures should probably be studied along with moisture, as Jones (1975) has done, more than they have been.

AERATION

The air available to a soil nematode is regulated greatly by pore space, soil depth, and temperature. In addition, the respiration rate of nematodes and other organisms is important in governing the amount of oxygen that will be available at any given time. Oxygen may be depleted faster than it can be replenished where there is an initial short supply, or when a large biomass rapidly consumes the oxygen available. Although oxygen requirements of nematodes vary considerably, no anaerobic plant nematodes are known. It would be surprising, indeed, to find any. Klekowski et al. (1972) assembled data for 68 soil nematode species and grouped them according to their known feeding habits. Species clustered into groups relative to respiration (Figure 7.4) and metabolic rates. Some groups overlapped whereas others did not. The predators clustered above the regression line, probably because of their frequently large size and greater activity. Some plant-parasitic

Table 7.2. Temperatures and nematode reactions

Nematode	Optimum temperature (C)	Comments	Reference
Aphelenchoides besseyi	About same at 23-30	16-35 C tested on Fusarium solani; generation time of 9-11 days	Huang et al., 1972
Aphelenchus avenae	30	Followed closely by 25 and 20 C; Rhizoctonia solani on agar	Evans and Fisher, 1970
Belonolaimus longicaudatus	25-30	20, 25, 30, 35 C tested	Robbins and Barker, 1974
Criconemoides curvatum	25	15-30 C tested	Malek and Jenkins, 1964
C. xenoplax	26, 22, 24	13-28 C tested	Seshadri, 1964
Ditylenchus dipsaci	18.3 (air)	18.3, 23.9 C tested	Barker and Sasser, 1959
D. dipsaci	15 (highest population peaks)		Tseng et al., 1968
D. dipsaci	15.6 (egg production)	11.1, 15.6, 22.8 C tested	Grundbacher and Stanford, 1962
D. myceliophagus	25.0		Evans and Fisher, 1969
Helicotylenchus dihystera	25 and 28.9	11.1-34.4 C tested	McGlohon et al., 1961
H. pseudorobustus	32.2 better than 23.9		Taylor, 1961
Hemicriconemoides chitwoodi	25	15, 20, 25, 30 C tested	Chang and Raski, 1972
Hemicycliophora arenaria	32.5	20-35 C tested	Van Gundy and Rackham, 1961
Heterodera avenae	0-7 followed by 15	Egg hatch	Fushtey and Johnson, 1966
H. betulae	28 (no reproduction at 31 and over)	14-35 C tested	Riggs et al., 1969
H. schachtii	27.5		Thomason and Fife, 1962
H. schachtii	20(?) penetration	15-35 C tested	R. N. Johnson and Viglierchio, 1969

Species	Temperature (C)	Process / comments	Reference
H. schachtii	27		Brzeski, 1970b
H. trifolii	25		Norton, 1967
Longidorus africanus	30–35		Lamberti, 1969
L. africanus	Symptoms 23.9 and above, but not at 12.8 and 18.3		Radewald et al., 1969
Meloidogyne arenaria thamesi	26.7 and 21.1	Hatching and reproduction 20, 25, 30 C tested	Thomason and Lear, 1961
M. graminis	28	Egg production 16, 22, 28, and 34 C on Zoysia grass	Grisham et al., 1974
M. hapla	21	Egg hatch	Wuest and Bloom, 1965
M. hapla	25	No reproduction at 15 and 35 C	Griffin and Jorgenson, 1969
M. hapla	30, 25		Irvine, 1966
M. hapla	25		Griffin, 1969
M. incognita	25	Egg hatch, 15–30 C tested	L. F. Johnson and Shamiyeh, 1968
M. javanica	24–28 (hatching)	18–22; 24–28; 28–32 C tested	Wallace, 1969
M. javanica	18–22; 24–28 (invasion number)	18–22; 24–28; 28–32 C tested	Wallace, 1969
M. javanica	18–22; 24–28 (invasion rate)	18–22; 24–28; 28–32 C tested	Wallace, 1969
M. javanica	About 15	Embryonic development	Wallace, 1971b
M. javanica	About 30	Hatching	Wallace, 1971b
M. javanica	Between 25 and 30	15–30 C tested; embryogenesis fastest at 30, but more eggs developed at 25	Bird, 1972
M. javanica	46 for 10 minutes	Embryogenesis and hatching suppressed	Bird, 1974
M. naasi	27	Reproduces from 16–32 C	Radewald et al., 1970
M. naasi	6–9 for 7 weeks followed by 21–24	Hatching	Watson and Lownsbery, 1970

Table 7.2. *(Continued)*

Nematode	Optimum temperature	Comments	Reference
M. ovalis	16	Maximum infection	Riffle and Kuntz, 1967
M. ovalis	20	Maximum development	Riffle and Kuntz, 1967
Merlinius joctus	30 better than 22		Norgren et al., 1968
Pratylenchus coffeae	33.3	Populations increased with increase from 23.9–33.3 C	Riggs et al., 1956
P. coffeae	29.5	24–35 tested	Radewald et al., 1971
P. neglectus	24	With *Verticillium dahliae*	Faulkner and Bolander, 1969
P. neglectus	30	Without *Verticillium dahliae*	Faulkner and Bolander, 1969
P. neglectus	At least 37.8		Mountain, 1954
P. neglectus	30	Invasion of corn	Townshend, 1972
P. penetrans	Greater at 30 than at 22.5 and 15		Patterson and Bergeson, 1967
P. penetrans	Greater at 16 and 24		Dickerson et al., 1964
P. penetrans	7 to 13	It takes fewer nematodes to cause injury to onion at 7–13 than 16–25 C	Ferris, 1970
P. penetrans	24	Reproduces 15–30 C; 15–33 C tested	Mamiya, 1971
P. penetrans	20	Invasion of corn	Townshend, 1972
P. scribneri	Maximum at 25 and 35		Thomason and O'Melia, 1962
P. vulnus	Disease worse at high temperatures tested (15.6–32.2)		Sher and Bell, 1965

Species	Temperature	Notes	Reference
Rotylenchulus parvus	30	20–35 C tested	Dasgupta and Raski, 1968
R. reniformis	29.5 (19 days)	15–36 C tested; life cycle completed within 27 days at 21.5 and 25 C	Rebois, 1973a
R. reniformis	27		Heald, 1975
Rotylenchus robustus	13 not as favorable as 18.5 and 24	21 and 27 C compared	Lear et al., 1969
Trichodorus christiei	30		Bird and Mai, 1967
T. christiei	25		Malek and Jenkins, 1964
Tylenchorhynchus claytoni	21.1–26.6	Wheat	Krusberg, 1959
	29.4–35.0	Tobacco	
T. dubius	20	Population increase	Brzeski, 1970b
Xiphinema americanum	20 or 24		Griffin and Barker, 1966
X. americanum	21		Lownsbery and Maggenti, 1963
X. bakeri	4 C at pF 2.4	Survival poorest at 30 C at pF 4.2	Sutherland and Sluggett, 1974
X. index	29.4		Radewald and Raski, 1962a
X. index, X. brevicolle, X. italiae, Longidorus brevicaudatus	28	16, 20, 24, 28 C tested	Cohn and Mordechai, 1970

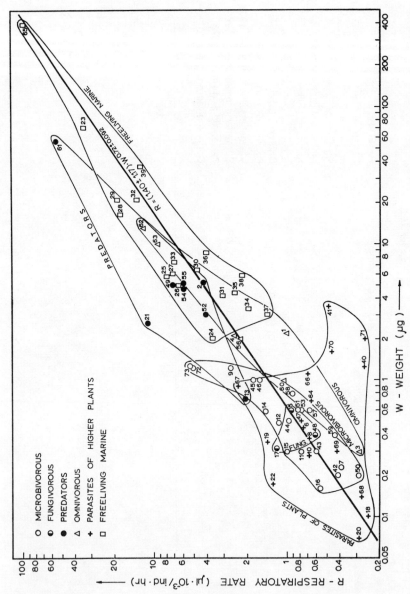

Figure 7.4. Respiratory rate of 68 nematode species: compilation of original and literature data. (From Klekowski et al., 1972. p. 399.)

nematodes had high respiration rates and others had low ones, a not very unexpected phenomenon in view of the heterogeneity within the group.

As expected, aeration can be important to a nematode from the beginning of life. The minimum oxygen concentration for the optimum hatch of *Aphelenchoides composticola* and *Ditylenchus myceliophagus* was 1.5 and 2.0%, respectively. Maximum juvenile development for both nematodes occurred at oxygen concentrations above 5% (Nikandrow and Blake, 1972). With normal oxygen concentration being about 20% in the soil, it is probable that hatch and development of these species are not often limited by the oxygen supply. Many nematodes, however, are certainly limited by oxygen during long periods in residual water. Wong and Mai (1973) found that when CO_2 was maintained at 0.03% the activities of *Meloidogyne hapla* were reduced at oxygen levels below 21% or at 40% compared with those at 21%. When oxygen was maintained at 21%, all levels of CO_2 above 0.03% were detrimental, except at 0.35%, where there was more invasion than at 0.03%. Hatch and movement occurred at 3.2% O_2, but there was no invasion.

Bhatt and Rohde (1970) observed that respiration of five species of nematodes generally was stimulated by the most dominating environmental factor of the natural habitat. Soil-inhabiting nematodes utilized O_2 most rapidly with high (1 to 2%) CO_2, whereas *Aphelenchoides ritzemabosi* did so with 0.03% CO_2. *Ditylenchus dipsaci*, *Pratylenchus penetrans*, and second-stage juveniles of *Anguina tritici* and *A. agrostis* respired from an osmotic pressure of 0 to 44.8 atmospheres. Respiration of their drought-resistant stages was stimulated by increasing osmotic pressure, which accompanies onset of drought. Supplemental Na^+ stimulated respiration of *A. tritici*, but K^+ did not.

pH

Although much is known about soil pH, less is known about its effects on nematodes. Statements are common that pH is of little consequence to soil nematodes, but the subject is far from closed. Changes in pH are so related to changes in other factors that it is difficult to separate the influences of one from the other. With this in mind, we will examine some correlations that have been made concerning nematode populations and pH with little reflection on cause and effect. Since pH is important biologically, it seems logical to expect that nematodes are affected by pH changes as are other forms of life.

Although nematodes, like many plants, seem to tolerate wide pH ranges, there are limits in which the hydrogen or hydroxyl ions, or other ions affected by pH, become toxic. It is important to remember when making judgments concerning pH or other factors and plant growth relative to nematodes that the hosts may be under stress and that nematode damage is enhanced in plants already stressed. Damage to a plant is the result of total stress, of which nematodes are only a part. Some correlations of nematode numbers or disease incidence for pH ranges are summarized in Table 7.3. It will be noted that there is diversity of findings in some instances. Lownsbery (1961), for example, found no significant difference in population levels of *Criconemoides xenoplax* around peach after 7 months at pH 5 and 7, but Seshadri (1964) found significantly more nematodes of the same species around grape after 3 months at pH 7 than at pH 5. In considering the data summarized in Table 7.3 or in discussing pH and nematode occurrence in general, it should be remembered that only certain pH ranges were studied and that if wider ranges had been incorporated, a normal distribution curve, however skewed, would probably be obtained as suggested by the data of Ellenby (1946) and Robinson and Neal (1956). Burns (1971) also found some semblance of this when she adjusted soil to pH 4.0, 6.0, and 8.0. pH 6.0 was the best for maintenance of several nematodes, including *Hoplolaimus galeatus, Pratylenchus alleni,* and *Xiphinema americanum.* She postulated that part of the explanation for lower maintenance at pH 4.0 could be due to the thicker epidermal wall of the roots and the higher levels of potassium, manganese, and phenols that occurred in plants at pH 4.0. Similar information was obtained with *Pratylenchus penetrans* when it was found that its reproduction after 30 weeks on "Vernal" alfalfa was significantly greater at pH 5.2 and 6.4 than at 4.4 and 7.3 (Willis, 1972). Forage yields at pH 4.4 were significantly less than at the higher pH values in both the infested and noninfested treatments, but the poor reproduction at pH 7.3 could not be attributed to insufficient substrate on which to feed. Work is needed to distinguish the correlations caused by pH differences from those caused by interrelated phenomena. If nothing more, however, correlations provide a locus around which to work.

FERTILITY

It is often stated that damage caused by nematodes can be overcome to various degrees by applications of fertilizers that stimulate root growth. Even though an increase in nematode density can result from such applications, the plant literally outgrows the damage caused by the nematode. If this is true, the converse is not always true; that is, reduced fertility does not always result in fewer nematodes. Dolliver (1961) found that moderately reduced fertilization of Wando peas resulted in more *Pratylenchus penetrans,* and Castaner (1963) found that significantly more *Helicotylenchus pseudorobustus* occurred in corn plots receiving no N-P-K than in plots receiving these elements at 179, 29, and 56 kg/ha, respectively.

Nitrogen

Nitrogen has two effects on nematode populations. Its application can result in nematode increase as with *Heterodera glycines,* presumably by providing more feeding sites through stimulation of root growth (Ross, 1959). Most reports, however, indicate that addition of nitrogen results in lower nematode numbers. Among them have been results with *Helicotylenchus pseudorobustus* (Castaner, 1966), *Hoplolaimus galeatus* (Castaner, 1966), *Pratylenchus minyus* (Kimpinski et al., 1976), *P. penetrans* (Patterson and Bergeson, 1967; Walker, 1971), *Heterodera glycines* on nonnodulating soybeans (Ross, 1969), *Belonolaimus longicaudatus* on turf (Heald and Burton, 1968), and *Meloidogyne incognita* on lima beans (Oteifa, 1955). It has been suggested that the ammonia produced was responsible for this phenomenon (Oteifa, 1955; Eno et al., 1955; Walker, 1971). Oteifa (1955) speculated that the ammonium ions were inhibitory to the hatching of *M. incognita* eggs because, after hatching and infection occurred, the nematodes developed normally.

Several forms of nitrogen have been employed and not all forms are equally effective. Walker (1971) found that nitrate was least effective for reducing numbers of *P. penetrans,* and Heald and Burton (1968) found that organic nitrogen in the form of activated sewage sludge reduced numbers of *B. longicaudatus* and *Hypsoperine graminis* on turf more than did inorganic nitrogen. Galling of tomatoes was reduced also by sludge additives, possibly resulting from toxicity of chemicals released during incubation of the sludge (Habicht, 1975). Unless phytotoxicity results, the decrease of nematodes with increasing amounts of nitrogen cannot be accounted for by less root tissue being available to the nematodes since, in most data where root weights were given, root weights increased with increasing nitrogen. Evidently other factors are operating directly on the nematode, or indirectly through the host. Nematodes in Manitoba soils generally decreased in clay with increasing nitrogen, but numbers increased in sand with increased nitrogen (Kimpinski and Welch, 1971), indicating that soil factors may be important.

Potassium

Although potassium is abundant in the soil, most of it is unavailable to plants. The addition of potassium fertilizers to soil can change the level of available potassium rapidly, but the effects may be short lived. Potassium is absorbed by plants in large quantities, and it leaches from the soil more readily than some other elements.

Kirkpatrick et al. (1964) found that the largest numbers of *Pratylenchus penetrans* in sour cherry orchards in New York State occurred under

Table 7.3. Correlations of nematode numbers or other phenomena with pH

Nematode	pH	Comments	Reference
		Positive correlation	
Criconemoides xenoplax	3.0, 5.0, 7.0	Grape; host growth poor at pH 3.0	Seshadri, 1964
Helicotylenchus platyurus	4.5–6.9	Iowa woodlands	Norton and Hoffmann, 1974
H. pseudorobustus	6.0–8.0	Iowa soybean fields	Norton et al., 1971
H. pseudorobustus	4.5–7.9	Iowa woodlands	Norton and Hoffmann, 1974
Heterodera avenae	About 5.0–7.5	Numbers in 625 field samples	Duggan, 1963
H. rostochiensis	Less than 3.0 to about 8.0	Hatch; peaked at 6.0	Ellenby, 1946
H. schachtii	5.0, 6.2, 7.2	Cabbage pot experiments	Brzeski, 1969
H. schachtii	4.5–7.2	Carrot	Brzeski, 1970a
Pratylenchus neglectus	4.5–7.2	Carrot	Brzeski, 1970a
P. penetrans	4.5–7.2	Carrot	Brzeski, 1970a
Tylenchorhynchus microdorus	4.5–7.2	Carrot	Brzeski, 1970a
T. nudus	5.0–6.5	Iowa prairie	Schmitt, 1969
T. maximus	5.0–6.5	Iowa prairie	Schmitt, 1969
Tylenchulus semipenetrans	6.0, 7.5	After 11 months	Van Gundy et al., 1964
Xiphinema americanum	4.5–7.4	Iowa woodlands	Norton and Hoffmann, 1974
		Negative correlation	
Heterodera rostochiensis	1.0–6.5	Hatch; best at 2.5	Robinson and Neal, 1956
H. rostochiensis (presumably)		Cysts in 53 peat soils	Smith, 1929
Hoplolaimus galeatus	6.0–8.0	Iowa soybean fields	Norton et al., 1971
Meloidogyne incognita acrita	4.5–6.5	Disease index	Kincaid et al., 1970

Organism	pH	Notes	Reference
Meloidogyne and *Pratylenchus*	4.5–6.2	Disease incidence	Kincaid and Gammon, 1957
Pratylenchus brachyurus	4.5–6.5	Disease index	Kincaid et al., 1970
P. crenatus	4.5–7.2	Carrot	Brzeski, 1970a
P. penetrans	4.5–6.5	Disease index	Kincaid et al., 1970
P. penetrans	4.8–7.5	Vetch	Morgan and MacLean, 1968
Tylenchinae-Psilenchinae	6.0–8.0	Iowa soybean fields	Norton et al., 1971
Tylenchorhynchus dubius	4.1–8.5	Cabbage	Brzeski and Dowe, 1969
T. dubius	4.5–7.2	Carrot	Brzeski, 1970a
T. nudus	6.0–8.0	Iowa soybean fields	Norton et al., 1971
T. vulgaris	5.5–9.5	Multiplication on maize	Upadhyay and Swarup, 1972
Tylenchus davainei	4.5–7.2	Carrot	Brzeski, 1970a
Xiphinema americanum	6.0–8.0	Iowa soybean fields	Norton et al., 1971
X. chambersi	4.5–6.4	Iowa woodlands	Norton and Hoffmann, 1974

No pH differences

Organism	pH	Notes	Reference
Criconemoides xenoplax	5.0–7.0	Peach, 7 months	Lownsbery, 1961
Heterodera rostochiensis	3.0–8.0	Cyst hatch, ?	Peters, 1926
H. rostochiensis	3.0–10.6	Hatch rate	Fenwick, 1951
Meloidogyne hapla	3.0–10.6	Not attractive	Bird, 1959
M. javanica	3.0–10.6	Not attractive	Bird, 1959
Meloidogyne sp.	4.0–8.5	Pineapple	Godfrey and Hagan, 1933
Tylenchorhynchus martini	3.5–7.5	Swarming	Hollis, 1958
Several		Sand dune	Yeates, 1968

141

potassium-deficient trees. Application of K_2SO_4 at 0 to 2.3 kg per tree, in the absence of phosphorus, increased the numbers of *Xiphinema americanum* and also increased plant top growth. When 2.3 to 4.5 kg K_2SO_4 was added per tree, the numbers of *X. americanum* decreased, but plant root and top growth also decreased. When deficient, optimum, and excessive amounts of potassium were applied to lima beans, only minor differences were noted in the development of *Meloidogyne incognita* between entry into the roots and development into fully grown females (Oteifa, 1953). As the potassium was increased, however, the development from fully grown females to maturity and egg deposition proceeded more rapidly. Oteifa (1951) also found that at a low inoculum level the reproduction rate apparently was limited by the available potassium. At higher inoculum levels and higher rates of potassium, the reproduction rate was correlated more with the amount of root tissue available and competition with other nematodes rather than with the level of potassium.

Other work also indicates that increasing potassium favors nematode or disease development. Kincaid et al. (1970) found that root knot and coarse root were positively correlated with the amount of K_2O applied, with maximum disease occurring at application of 896 kg/ha. Using invaded cucumbers in quartz sand, Marks and Sayre (1964) found that low potassium levels retarded the rate of *M. incognita* development, but development was accelerated at higher levels of potassium. The final populations in the roots were significantly increased with high levels of potassium. *Meloidogyne javanica* and *M. hapla* were not influenced by potassium, however. Oteifa (1952) found that increased potassium overcame the damaging effects of *M. incognita* in lima beans. Later, Oteifa et al. (1964) reported that potassium initially increased the numbers of *Tylenchorhynchus latus* around cotton, but this did not continue beyond the early stages of cotton growth. Unfertilized plants were generally least tolerant to the nematode despite the lower levels of infestation.

Thus, in general, any increase in potassium from low to optimum or nontoxic levels increases the development or numbers of the nematode, or the disease incidence. There are exceptions, however, and further work is necessary.

Phosphorus

As is well known, phosphorus is important to plants. Since its availability and chemistry in the soil is complex, other factors must be taken into account when studying the effect of the element on nematode populations, directly or indirectly. Kirkpatrick et al. (1964) found that numbers of *Xiphinema americanum* increased around sour cherries with an increase of phosphorus and no addition of potassium. Yields of oats infested with *Heterodera avenae* declined rapidly in soil containing less than 8 mg phosphorus per 100 g soil (Fidler and Bevan, 1963). Crop failures were obtained when nematode levels were at 18 eggs/ml and phosphorus at 5 mg/100 g. In a study of nematodes in pastures in New Zealand, Yeates (1976) recorded a fourfold increase of *Pratylenchus* with 500 kg superphosphate per hectare compared with 125 kg/ha. Similar increases were obtained with applications of lime compared with no lime. There was little increase in plant dry matter despite the increase in soil fertility.

Calcium

As with some other elements, there is little information on the direct effect of calcium on nematodes in the soil. Even disregarding the indirect effects of calcium through plant growth, it is still difficult to evaluate the effects of calcium since most additions of calcium to the soil also increase the pH. Kincaid et al. (1970) found that root knot and coarse root varied with the amount of CaO applied to the soil. The numbers of *Radopholus similis* recovered from roots of citrus were reduced after heavy applications of lime, but plant growth was adversely affected by the large amounts of lime (Tarjan, 1961). Work by Yeates (1976) was mentioned under phosphorus.

ORGANIC MATTER

Ever since Linford et al. (1938) reported reduced galling after chopped plants were incorporated into root-knot-infested soil, there has been great interest in the use of organic matter as a means of nematode control. The authors pointed out that decomposition of plants probably results in increased total nematode numbers, which in turn support animal and plant life destructive to harmful species. While the parasitism and predation that result from organic amendments are undoubtedly important in controlling nematode populations (see Chapter 9), greater attention has been made in recent years to other explanations as contributing to control.

Organic Matter Amendments

A number of the organic amendments have been incorporated into the soil as green manure. These include ground tobacco stems (Miller et al., 1968), alfalfa pellets or hay (Mankau, 1968; Mankau and Minteer, 1962), lespedeza hay or oat straw, flax hay (L. F. Johnson et al., 1967). peat moss (O'Bannon, 1968), and plant products such as sawdust (Singh et al., 1967), paper, cottonseed meal (Miller et al., 1968), hardwood bark (Malek and

Gartner, 1975), cornmeal, soybean meal and oils (Walker, 1969; Walker et al., 1967), oil cakes (Singh and Sitaramaiah, 1966), cotton waste, sugar beet pulp, castor pomace (Lear, 1959; Mankau and Minteer, 1962), chitin, cellulose, and mycelial residues (Miller et al., 1973), animal manure or products (Bergeson et al., 1970; Mankau and Das, 1969; Mankau and Minteer, 1962), and other organic amendments (Heald and Burton, 1968). Mankau and Minteer (1962) tested eight organic substances and found that only steer manure failed to reduce the numbers of *Tylenchulus semipenetrans* juveniles substantially in 84 days. Castor pomace eliminated all juveniles but did not produce a substance toxic to nematodes. Although incorporation of green manures into the soil has been effective in reducing disease, Johnson et al. (1967) found, in tests with *M. incognita,* that nematicides were generally superior to crop residues. Mankau (1968) concluded that the number of juveniles of *M. incognita* remained about the same with alfalfa or steer manure amendments as with inorganic amendments, but the infectivity and survival was reduced in the organic amendments. *Actinomycetes* increased after chitin amendments, and this was associated with a decrease of secondary invasion of the root-knot galls in okra (Bergeson et al., 1970). Plants growing in chitin-amended soil were larger and the roots were less necrotic than those growing in nonamended soil even though the degrees of galling were similar.

Generally, organic amendments have given enough encouragement with several crops and nematodes to continue further work. Often, however, the quantity of organic amendments needed to realize effective nematode control exceeds practical supply. Although there is promise of nematode control by organic amendments, much work is necessary before practical implementation is a widespread reality. Mechanisms of control have not been worked out in most instances. It has been demonstrated that environmental factors such as temperature, moisture, and pH are important relative to the degree of control obtained (Johnson, 1962), but this will have to be manipulated for each instance.

In some of the aforementioned accounts, a period of incubation was required to obtain the best control from organic amendments. The incubation period differed with the material being added to the field. Material that was primarily cellulose required more incubation than herbaceous matter. Incubation periods up to one year were required for maximum nematode reduction with highly cellulosic materials, but little or no benefit was gained by incubations of less than 2 months (Miller et al., 1968). Where there was little incubation time, cottonseed meal proved more practical, as 100% control occurred in 3 months. Using large amounts of chopped tobacco stems, Johnson (1959) also noted that a 30-week incubation was necessary to reduce infection. Variation of the incubation temperature had different results with different types of organic material (Walker et al., 1967). Materials that were most easily broken down seemed most readily affected.

Organic Matter Naturally in Soil

Although the relationship of organic matter and nematode populations has been studied widely with respect to the addition of organic matter to soil, fewer observations have been reported on natural associations of organic matter and nematodes. Since organic matter will influence pH, cation exchange capacity, moisture, and microbial populations, among other properties, these complexities must always be considered. Mineral soils usually have an organic matter content of less than 7%, but other soils may range up to 100% organic matter in pure peat. Both positive and negative correlations of nematodes with organic matter have been observed, but the paucity of information leaves much to be desired.

Psilenchus and *Aphelenchoides* were found more frequently or in larger numbers in Maryland soils with relatively higher organic matter (Jenkins et al., 1957), and *Dolichodorus heterocephalus* was found in poorly drained New Jersey soils that were high in organic matter (Hutchinson et al., 1961). Brzeski (1965) reported that *Hemicycliophora parvana, Hemicriconemoides wessoni,* and *Criconemoides ornatum* in Florida citrus groves occurred in soils with relatively more organic matter. Marchant (1934) sampled 29 soils in Manitoba that were from 2.2 to 90.1% organic matter and found a positive correlation of total nematodes with organic matter irrespective of soil type. Positive correlations of organic matter and members of the Tylenchinae, Psilenchinae, and nonstylet nematodes were found in soybean fields in a normal rainfall year; negative correlations with organic matter were found with the dorylaimids, all stylet nematodes, and *Helicotylenchus pseudorobustus*. Only dorylaimids were significantly correlated, again negatively in a dry year (Norton et al., 1971). Oteifa and Abdelhalim (1957) reported a positive correlation of organic matter with total nematodes in Egyptian soils ranging from 0.6 to 2.2% organic matter, but these occurred at different depths down to 15 inches.

Mechanisms of Control

Although there has been great interest in the practical results of adding organic matter for nematode control, there is increasing interest in the mechanisms of this control. Much of this work has been based upon the association, correlations, and isolation of soil chemicals that appear to be toxic to nematodes.

An early indication that organic decomposition products were toxic to nematodes was expressed by Thorne (1926a), who indicated that heat and certain nematotoxic gases were produced during the decomposition of sweet clover. Many other materials have since been demonstrated to have toxic effects during their decay. Extracts of municipal waste reduced motility of *Belonolaimus longicaudatus* in 3 hours or killed an entire population after 144 hours immersion in the extract (Hunt et al., 1973). Taylor and Murant (1966) cite the degradation of tannins released by raspberry canes as a possible mechanism for the reduction of *Longidorus elongatus*. It is known that tannins degrade into nematotoxic polyphenols. Singh and Sitaramaiah (1966) speculated that the inability of *Meloidogyne javanica* to exude eggs, and the inability of previously released eggs to hatch, was the result of a toxin in the soil solution. This toxin was purported to be a degradation product of the oil cakes used for soil amendment. Extracts of decomposing plant residues from rye were selectively nematicidal to *Meloidogyne incognita* and *Pratylenchus penetrans*, but it took higher concentrations to immobilize *Rhabditis, Cephalobus,* or *Plectus* (Patrick et al., 1965). Extracts from decomposing rye and timothy were selectively nematicidal to the same parasitic nematodes under laboratory and field conditions (Sayre et al., 1965). The extracts were not toxic to saprobic nematodes when used at the same concentrations that immobilized parasitic ones. One toxic agent was identified as butyric acid. The chemical and nematicidal activity of chemically pure butyic acid was identical to that isolated from decomposing debris. Volatile fatty acids were present in decaying residues in greater quantity in the laboratory than in the field. Butyric acid and the residue extracts were active only at pH 4.0 to 5.3. Similarly, L. F. Johnson (1974) found that extracts from oat straw and flax leaves and stems were toxic to eggs and juveniles of *M. incognita* in vitro, but to a lesser degree in soil amended with similar amounts of the plant material. It also has been found that rapidly decomposing organic matter in cornmeal-amended rice soil caused a rapid kill of nematodes (Hollis and Rodriguez-Kabana, 1966). N-Butyric acid and smaller quantities of propionic acid increased rapidly in cornmeal-treated soil (Hollis, 1958). Concentrations of butyric acid comparable to those found in the treatments killed all nematodes within a few hours, and nontoxic levels of propionic acid had an additive effect. These acids are produced by *Clostridium butyricum*, which was present in the tests. This was not solely a pH effect since acetic acid was the main organic acid found, and it did not have any nematicidal effect at the concentrations found. *Tylenchorhynchus martini,* which was the main nematode studied in these tests, was unimpaired at pH 3.5.

Stephenson (1945) subjected *Rhabditis terrestris* to solutions of 16 different organic or inorganic acids, all of which produced a pH of 2.8 or under. The mineral acids had a greater effect than the organic acids. Nitric acid killed the fastest at comparable normalities, killing in about 1½ minutes; succinic acid was slowest to kill, taking over 16 minutes. The general effects of the acids were to increase the stickiness of the cuticle and to cause a bulge in the cuticle in the anterior end of the body. The writer concluded that the effects were due partly to lowered pH and partly to the dissimilar effects of undissociated acids. Both Stephenson (1945) and Johnston (1959), the latter author working with *Tylenchorhynchus martini*, found that formic, acetic, propionic, and butyric acids immobilized nematodes. Using a *Dorylaimus* species, Banage and Visser (1965) tested the effects of formic, acetic, propionic, butyric, and valeric acids in distilled water at concentrations ranging from 1 to 0.001 M and found that the solutions were exponentially toxic to the nematode. The acids were not toxic at 0.0001 M. The sodium salts were less toxic than the acids themselves. There was also evidence that the undissociated acid molecule was the chief factor. Formic acid was less toxic at any value and the others were about equal.

Ponchillia (1972) added from 4 to 12% peat to two mineral soils and found that survival and migration of *Xiphinema americanum* was significantly lowered by adding peat. He extracted humic acid and fulvic acid fractions from peat and found that the fulvic acid fraction was deleterious to the nematode. Later, Elmiligy and Norton (1973) found that reproduction of *Aphelenchoides goodeyi* and *Helicotylenchus pseudorobustus* was reduced when the nematodes were treated with fulvic acid, as compared with water treatments, sodium humate, or humic acid.

It seems that the effects of organic amendments or organic matter in the soil are due to parasitism and predation, the liberation of toxins directly from the organic matter or their decomposition products, or possibly competition as a result of gas exchanges. This last aspect has been touched only a little as it bears on this problem.

OSMOTIC PRESSURE

Osmosis is defined as the diffusion of a liquid through a differentially permeable membrane into a solution in which the solvent is the same liquid or into another liquid with which it is miscible. Osmotic pressure, measured in atmospheres, is the maximum pressure that can develop in a solution separated from pure water by a rigid membrane permeable only to water. Since it is difficult, and often impractical, to determine osmotic pressures directly (manometrically), they are usually determined indirectly from the lowering of the vapor pressure or elevation of the boiling point, or cryoscopically, since osmotic pressure is related to all of these. We speak of the osmotic pressure of a solution irrespective of the subjection of the solution to the necessary conditions to attain that maximum pressure. The osmotic pressures of many solutions have been calculated and may be found in chemical handbooks. In plants, the root surface acts as a membrane; in nematodes, the body wall acts as a membrane.

As we will see, soil and plant nematodes can withstand pressure close to 10 atmospheres, sometimes much more, before their mobility and survival are greatly affected. Since the osmotic pressure in the soil rarely exceeds 2 atmospheres in humid regions, osmotic pressure probably is not a major force in the infection processes of nematodes. Osmotic pressure studies are valuable, however, even if the pressures used exceed those normally found in the soil, since they give insight into the permeability of nematodes and eggs to chemicals that might affect their survival or be effective in their control. It is impossible to separate salinity from changes in osmotic pressure since any change in salinity will change the osmotic pressure. Thus they both are discussed together.

Most soil environments are dynamic, and only those nematodes capable of making the necessary adjustments flourish or survive in a changing habitat. Beyond certain limits, adjustment is not possible and the individual, and perhaps the population, dies. One of the most fluctuating properties of any soils is its moisture content and the concomitant changes in suction and osmotic pressure. Different nematodes have different tolerances and optima for various activities and during various phases of their life cycle. These in turn affect the distribution and population size of nematodes and may be limiting or preferential. Although much work is necessary concerning factors favoring or limiting nematode estabishment, it seems that the most nematode distribution around crop plants is not greatly affected by the salinity or osmotic pressure changes commonly associated with that crop. A balance has already been reached or there would be no association. The fact that some nematodes occur in habitats of higher osmotic pressure and suction than are normally encountered while other species are absent suggests that differences exist in nematode tolerances. For example, Rau (1963) reported that *Belonolaimus maritimus* seemed more tolerant of salt or brackish water than *B. longicaudatus*. *Meloidogyne spartinae* produces galls on *Spartina alterniflora,* which is a salt-water tolerant plant of tidal marshes (Fassuliotis and Rau, 1966). Egg hatch of *M. spartinae* was significantly greater in salt solutions of 0.2, 0.3, and 0.4 M than in solutions of 0.0, 0.1, and 1.0 M and in marsh water controls (Fassuliotis and Rau, 1966). Eggs did not differentiate at 0.0 M, and development did not proceed beyond the blastula stage at 1.0 M. Members of the Dorylaimida, i.e., *Trichodorus pachydermus, T. similis,* and *Longidorus elongatus,* were more susceptible to increasing desiccation of the soil and higher osmotic pressures than members of the Tylenchida, i.e., *Tylenchorhynchus dubius,*

Pratylenchus penetrans, and *Rotylenchus robustus* (Wyss, 1970b). *Tylenchorhynchus dubius,* which was found in the uppermost soil layers, was the most tolerant of drying and high osmotic pressures, while *Trichodorus* spp. occurred in the deeper soil layers and were the most susceptible.

Most nematode work in saline soils in the United States has been done in areas where saline, sodic, or saline-sodic soils are important in agriculture. Machmer (1958) found that *Tylenchulus semipenetrans* and *Meloidogyne incognita acrita* tolerated different soil salinity levels around host plants in the lower Rio Grande Valley in Texas and survived, although they were impaired. More citrus nematodes were recovered from roots subjected to higher than lower salinity levels. Van Gundy and Martin (1961) stated that citrus growing under marginal conditions, where the soil contained excess $CaCO_3$ or was high in exchangeable sodium, had moderate growth when not infected with *T. semipenetrans* but poor growth in the presence of the nematode. Working with *Rotylenchulus reniformis* in Texas, Heald and Heilman (1971) found that the nematode occurred equally in fields with high or low salinity. In greenhouse tests, the nematode significantly damaged cotton at salinity levels of 2, 6, 12, and 18 mmho/cm, and nematode injury increased as soil salinity increased. $CaCl_2$ and NaCl were the predominant salts. Dry weights of tops of tomatoes infected with *Meloidogyne javanica* were significantly less at a conductivity of 4 mmho/cm than at 8 mmho/cm, indicating that there is an interaction within the plant between nematode infection and salinization, with its associated changes in osmotic pressure (Maggenti and Hardan, 1973).

Dropkin et al. (1958) subjected eggs of *Heterodera rostochiensis* and *Meloidogyne arenaria* to molarities of 0.1, 0.2, 0.3, 0.4, and 1.0 M NaCl, KCl, $CaCl_2$, $Ca(NO_2)_2$, NH_4NO_3, and dextrose. Although there was some variability with the chemicals used, in general there was marked inhibition of hatching in solutions between 0.2 and 0.4 M and above. The osmotic pressures of the solutions that caused marked inhibition of hatching were considerably above those found in the soil, but the authors pointed out that osmotic concentrations in cysts and egg masses possibly may reach such levels and thus affect survival. Many of the eggs were capable of hatching after removal from the test solutions to water.

Depending on the osmotic pressure of the solution, nematodes will lose their mobility when subjected to various hypertonic solutions from 0 to 80 atmospheres. This is not the same for all nematodes, as was found when at least six different species were subjected to hypertonic solutions of urea and to NaCl (Viglierchio et al., 1969). The relative nematode response in order of increasing tolerance was: *Rhabditis* spp., *Pratylenchus vulnus,* *Hemicycliophora arenaria, Tylenchulus semipenetrans, Meloidogyne hapla,* and *Ditylenchus dipsaci* (Figure 7.5). The electrolytes also differed at equal

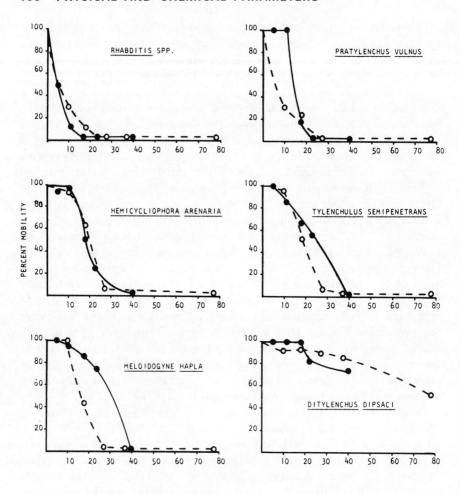

OSMOTIC PRESSURE (ATM)

Figure 7.5.   The mobility of nematodes after 24 hr in hypertonic solutions, followed by transfer to distilled water. Percent mobility calculated with respect to water controls (From Viglierchio et al., 1969, p. 16.)

osmotic pressures; for increasing mortality of *D. dipsaci* they were NaCl, Na_2SO_4, RbCl, and KI; for *Rhabditus* spp. they were Na_2SO_4, RbCl, NaCl, and KI. Mobility was not affected greatly, if at all, until 10 atmospheres or more was reached. Since the osmotic pressure in agricultural soils at field capacity rarely exceeds 2 atmospheres where water is not limiting, the data cannot be applied readily to some fields. Soil at the wilting range of plants is around 15 atmospheres, however, and thus the data may be applicable in these instances. Viglierchio et al. (1969) speculated that there are three principal ways that the osmotic treatments injured nematodes: (1) irreversible dehydration of tissue, (2) toxicity of chemicals in solution, and (3) excessive uptake of distilled water, after regulation in hypertonic solutions, resulting in body rupture.

Also using *D. dipsaci,* Blake (1961) found that movement of the nematode in 250 to 500 μm sand was the same whether the sand was wetted with distilled water or 0.1 M urea. Movement was fastest in both treatments when the sand pores began to drain. This indicates that movement is dependent only on moisture content and is independent of osmotic potential in the range used. Movement of *D. dipsaci* was not affected by sand saturated with solutions ranging from 0.0001 to 0.3 M urea and drained by suction of 32 cm water. Movement decreased at 0.3 M (pF = 3.87) and ceased at 1 M (pF = 4.35). Thus one can see readily the impracticality of large-scale application of sugar to the soil for control of nematodes, as has sometimes been suggested.

Also using a particle size of 250 to 500 μm saturated with urea at 10^{-3}, 10^{-2}, 10^{-1}, and 1 M, and distilled water with suction adjusted to 30 cm water, Wallace and Greet (1964) found that the mobility of *Tylenchorhynchus icarus* was about the same in water, 10^{-3}, and 10^{-2} urea, but mobility was less at 10^{-1} and was greatly inhibited at 1.0 M. The nematodes became shriveled at 1.0 M, indicating water loss. The data indicate that *T. icarus* could survive for a time in dry soil. Roggen (1966) has shown that the osmotic pressure of *Xiphinema index* can change when the nematode is subjected to sucrose solutions of different molarity. He also found that *X. index* containing the grape fanleaf virus had a higher internal osmotic pressure than nematodes without the virus.

Anguina tritici, A. agrostis, Ditylenchus dipsaci, and *Pratylenchus penetrans* respired well at osmotic pressures ranging from 0.224 to 44.8 atmospheres, but *P. penetrans*, which is the most susceptible to desiccation, respired the least at the high osmotic pressures (Bhatt and Rohde, 1970). *Anguina tritici* and *A. agrostis*, which can withstand periods of desiccation the longest, respired more than the others at high osmotic pressures. *Ditylenchus dipsaci,* which is intermediate in ability to withstand desiccation, was intermediate in respiration at 44.8 atmospheres, but not greatly so compared with *P. penetrans*. Since the osmotic pressures of plant cells commonly are within 10 to 20 atmospheres and range considerably to either side, such correlations may have merit.

OXIDATION-REDUCTION POTENTIAL

Little work has been done relating the oxidation-reduction potential of soils to nematode occurrence. This is unfortunate, because the oxidation-reduction potential varies so much in the soil that it possibly could tell us much directly or indirectly about nematode occurrence and populations. Since oxidation-reduction is based upon transfer of electrons that may or may not involve oxygen, the potential is not always a reflection of aeration. Free oxygen in the soil water greatly modifies the redox potential and thus may reflect aeration, however. The amount of oxidation-reduction occurring in the soil determines to a great extent the microbial activity, which in turn influences many aspects of the soil environment. It is not known how much the potential per se affects nematode activity, if any, but indirectly it could contribute much and merits further investigation. In soils, the oxidation-reduction potential generally decreases with increasing pH. Flooded soils generally have a low oxidation-reduction potential although this can be modified greatly by organic matter and minerals. As with most complex systems, it is difficult to separate out the limiting or controlling factor. For example, Hollis and Rodriguez-Kabana (1966) associated a reduction of nematodes in rice soil supplemented with cornmeal. They speculated that nematode control may be due in part to the increase of anaerobic bacteria, which are favored by low redox potentials. These bacteria are known to produce organic acids and other substances that have been shown to be toxic to nematodes. These reactions along with greatly reduced amounts of oxygen may all contribute to the reduction in the numbers of nematodes.

The redox potential has been suggested as a governing factor in nematode attraction. Bird (1959), in referring to Linford's work in which root-knot juveniles were attracted to decaying tissues, speculated that the nematodes moved toward an area of lower redox potential caused by rapidly multiplying bacteria. Bird also speculated that attraction of juveniles to anaerobic yeast colonies on agar was due to movement toward lower potentials, since yeasts will lower the redox potential more rapidly under anaerobic than aerobic conditions. Klingler's (1961) work did not confirm this hypothesis. He passed substances through fine canulae into agar and measured the attraction or repulsion of *Ditylenchus dipsaci*. Carbon dioxide attracted the nematodes, but oxygen, lowered the redox potential but unlike hydrogen, was attractive. Potassium permanganate, an oxidizing agent, raised the redox potential but was attractive. Obviously, much work is needed on oxidation-reduction and the accompanying complexities and their relation to nematodes.

SURVIVAL

Different stages of development have different
requirements and tolerances.

A nematode's ability to survive adverse environments is affected by all aspects of its physiological and morphological makeup. Most plant nematodes probably are subjected to unfavorable conditions sometime during their lives, and the persistence of a species in a habitat indicates that survival mechanisms are operative. Naturally, if one or a series of adverse conditions persists, the population or species will diminish or perish in that habitat. This is put to practical use agriculturally by practices such as crop rotation in which a suitable food source is removed. If deprivation of food is coupled with unfavorable edaphic or climatic conditions, the period necessary for control or eradication can be shortened. An adverse environment every few years is possibly sufficient to keep a population at moderate to low levels. This is probably true in the northern latitudes where some nematode species seldom increase to the damaging proportions that they attain in warmer climates. Although this might be due partly to the shorter growing season in northern latitudes with less time for progeny production, and thus cannot exactly come under the category of survival, it is nevertheless just as important in keeping populations restricted as partial kill due to adverse local environments.

LENGTH OF NEMATODE SURVIVAL ——————————————————————

The periods of nematode survival as listed in Table 8.1 should not be regarded literally in most instances. Many results derived from laboratory

Table 8.1. Length of nematode survival[a]

Nematode	Length of survival	Comments	Reference
Anguina agrostis	4 years	Galls, room temperature	Fielding, 1951
A. amsinckia	4 years, 4 months	Cool, dry	Steiner and Scott, 1934
A. balsamophila	24 years	Plant tissue, room temperature	Fielding, 1951
A. tritici	28 years	Galls, room temperature	Fielding, 1951
A. tritici	32 years	5 C	Limber, 1973
Aphelenchoides subtenuis	3 years, 2 months	Infected plants on open shelf	Christie, 1959
A. xylophilus	1 but not 2 years	In blue stain wood	Steiner and Buhrer, 1934
Criconemoides xenoplax	2 years	Flooded soil (low oxygen)	Bird and Jenkins, 1965
Ditylenchus dipsaci	23 years	Plant tissue, room temperature	Fielding, 1951
D. dipsaci	2 years	Nonhost, field soil	Lewis and Mai, 1960
D. dipsaci	242 days	Without host, 15 and 21 C	Miyagawa and Lear, 1970
D. dipsaci	212 days	Juveniles infective, soil at 15 C	Miyagawa and Lear, 1970
D. triformis	2½ years	Dry storage, 24–26 C	Hirschmann, 1962
Helicotylenchus dihystera	250 days	Air dry soil, room temperature	McGlohon et al., 1961
Hemicycliophora arenaria	6 months	Adults, fallow soil, 25 and 30 C	Van Gundy and Rackham, 1961
Heterodera avenae	5.5 years	Eggs, 40% relative humidity at 15 C; 75% relative humidity at 5 C	Meagher, 1974
H. glycines	630 days	Water (juveniles), 0–12 C	Slack et al., 1972
H. glycines	84 months	Infested soil (moist), 8–20 C	Slack et al., 1972
	72 months	Infested dry soil, 8–20 C	
	90 months	In clay at fluctuating moisture	
H. glycines	8 months	Soil peds	Epps, 1969

154

Species	Survival[a]	Condition	Reference
H. rostochiensis	2½ years	Cysts, room temperature	Ellenby, 1955
H. schachtii	3 months	Encysted eggs, fallow soil to 52 C	Thomason and Fife, 1962
Hoplolaimus galeatus	2–4 days	Air dry soil, room temperature	McGlohon et al., 1961
Meloidogyne incognita	Up to 30 days	Hanging drop slides	Shepperson and Jordan, 1974
M. naasi	7 weeks	Survived well at 16 C; less at other temperatures	Radewald et al., 1970
Paratylenchus dianthus	4 years, 7 months	Preadults, moist soil	Rhoades and Linford, 1961b
P. projectus	70 days	Air dry soil, room temperature	McGlohon et al., 1961
Pratylenchus penetrans	11 months	Air dried in greenhouse	Rössner, 1971
P. thornei	50 weeks	Clay-loam soil, 10 or 20 C	Baxter and Blake, 1968
P. thornei	2 weeks	Clay-loam soil, 40 C	Baxter and Blake, 1968
Radopholus similis	Less than 6 months	Without host in field soil	Tarjan, 1960
Sphaeronema californicum	At least one year	In soil with roots at 5 C	Raski and Sher, 1952
Trichodorus christiei	2 to 4 days	Air dry soil, room temperature	McGlohon et al., 1961
Tylenchorhynchus claytoni	170 days	Air dry soil, room temperature	McGlohon et al., 1961
Tylenchus polyhypnus	39 years	Rye, herbarium	Steiner and Albin, 1946
Xiphinema americanum	116 days	Field, soil 0 C or below	Norton, 1963
X. americanum	49 weeks	Soil at 10 C	Bergeson et al., 1964
X. americanum	9 months	Moist sand at 8 C	McGuire, 1973
X. bakeri	6 months	Fallow soil at 0–30 C	Sutherland and Ross, 1971
X. bakeri	32 weeks	Natural soil in polyethylene bags, 4 C and pF 2.4	Sutherland and Sluggett, 1974
X. index	6 months	Many survived in soil	Taylor and Raski, 1964
X. index	7 to 9 months	Few survived in soil	

[a] Indicates longest known period, either by actual survival or by length of experiment.

and greenhouse experiments reflect "the conditions of the experiment," which often vary considerably from conditions encountered in more natural situations. Despite environmental differences between greenhouse or growth chambers and those the field, it is probable that differences in results of greenhouse and field research usually are more magnitudinal than directional. Data such as those accumulated in Table 8.1 may indicate that certain nematodes have a greater or lesser survival capacity than others. Our ignorance on the biology of most nematodes makes such information by no means definitive. In some instances, such information may reflect extremes and not norms. Just as survival will vary with the stage of nematode development, survival of the same stages will also vary with the environment. An extreme example is *Anguina tritici*, which can survive for up to 32 years in galls. Juveniles of the same species are not noted for their survival capacity in the field, however, and one-year crop rotation usually will control the disease.

Van Gundy (1965) used the terminology of Keilen in discussing hypobiosis in nematodes. As Van Gundy pointed out, the metabolic level sometimes is not known or is difficult to ascertain, and it is frequently difficult to categorize individuals. Cryptobiosis is usually the result of dehydration or low temperatures in which all reversible life processes are suspended for considerable time. Dormancy is quiescence in which there is reduced but measurable metabolism and is limited usually by metabolic exhaustion or extreme environment. If the nematodes are not killed, activity usually resumes rapidly upon return to favorable environments. Dormancy is subdivided into cryobiosis, anhydrobiosis, osmobiosis, and anoxybiosis when the limiting factor is mainly cold, desiccation, high solute concentration, or low oxygen, respectively. Since some properties are not independent of each other, more than one factor may be operative. Inherent factors involved in survival include the amount of fats and carbohydrates in the nematode, including eggs, the rate of respiration, the structure and chemical composition that governs permeability of the cuticle, and the reproductive capacity of the nematode. Physiological aspects of survival have been reviewed and discussed by Van Gundy (1965).

EFFECT OF MOISTURE ON SURVIVAL

Nematode survival depends on many factors, but the absence of moisture is probably the most important. There are instances, however, in which nematodes can survive dry conditions for many years (Table 8.1), and there is no doubt that there is a wide spectrum of requirements and tolerances for moisture. Nematodes such as some species of *Ditylenchus* and

Aphelenchoides that inhabit stems, leaves, or flowers are especially capable of surviving long drought. Other species within these genera are not known for their long survival, although perhaps sufficient information is lacking. *Ditylenchus dipsaci* is resistant to drying, but *D. destructor* is very susceptible and *D. myceliophagus* is intermediate (Goodey, 1958). Individuals of *Aphelenchoides* can survive for several months (Table 8.1), and Franklin (1955) speculates that *A. parietinus* can withstand considerable desiccation. Most soil nematodes, except those in the cyst state, probably survive for much shorter periods even under the best of conditions.

It is reasonable to expect that some nematodes survive better under some moisture conditions than others. *Tylenchorhynchus martini*, for example, survived best in soil between 40 and 60% of field capacity; survival was lowest at 11% and at saturation (Johnston, 1957). All *Hoplolaimus galeatus* and *Trichodorus christiei* were killed within 2 to 4 days in an air dry soil at room temperature, while 14% of *Helicotylenchus dihystera* survived at least 250 days in the same soil. Individuals of *Tylenchorhynchus claytoni* and *Paratylenchus projectus* survived 170 and 70 days, respectively (McGlohon et al., 1961). *Trichodorus* was more susceptible to desiccation than *Pratylenchus penetrans* and *Rotylenchus robustus* (Rössner, 1971).

Kable and Mai (1968a) maintained soils at a relatively constant pF value and found that survival of *Pratylenchus penetrans* increased as the soil moisture tension increased between pF 0.0 and 4.2, but found that the degree of survival varied in different soils (Figure 8.1). Townshend and Webber (1971) reported that *P. penetrans* survived more in soils with high rather than low moisture tensions. Immobilization in thin moisture films as indicated earlier by Wallace and good aeration probably account for greater survival in drier soil. The interaction of temperature and moisture also affects survival. Survival of *P. minyus* was greatest at low temperatures provided it was above freezing (Townshend, 1973). At low temperature and high moisture tension, the nematodes have low metabolism, thus depleting food reserves more slowly. Low moisture tensions increase nematode mortality. It has also been found that *Anguina tritici* survives adverse temperatures better in dry than in wet galls (Bloom, 1963).

Rate of Water Loss

Not only is the amount of water lost important in nematode survival, but the rate of water loss is also a factor (Simons, 1973). Second-stage juveniles of *Heterodera rostochiensis* and *H. schachtii* are vulnerable to desiccation, but the latter is more so. This is possibly because juveniles of *H. schachtii* lose water at a significantly higher rate than do those of *H. rostochiensis*

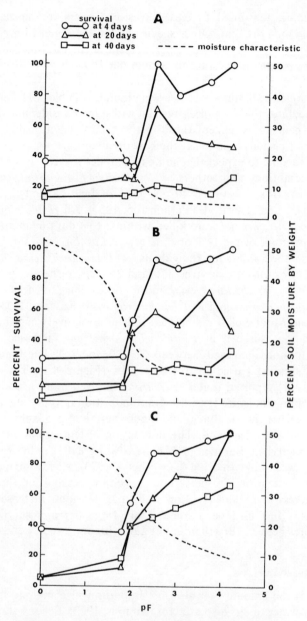

Figure 8.1. Survival of *Pratylenchus penetrans* at 20 in. Colonie sandy loam soil. (A), Dunkirk sandy loam soil (B), and Lockport clay loam soil (C), at moisture tensions of pF 0 to 4.2. (From Kable and Mai, 1968a, p. 108.)

(Ellenby, 1968). Viglierchio (1961b), who found that rapid drying or moderately slow freezing decreased the hatch of *H. schachtii*, speculated that the effect may be a function of the rate of drying and not the degree of dryness. Likewise, Huang and Huang (1974) found that survival of *Aphelenchoides besseyi* was inversely related to the extent and rate of dehydration.

Stage of Nematode Development

Different developmental stages of nematodes are known to differ in their susceptibility to desiccation. The egg often is considered to be one of the more resistant stages, and there is evidence that preadults of some species are more resistant to desiccation than the adult or other juvenile stages. Probably the best known example of the latter is *Ditylenchus dipsaci*, but the phenomenon has also been shown for *Paratylenchus projectus* and *P. dianthus* (Rhoades and Linford, 1961b). Young nematodes of *Xiphinema bakeri* are more susceptible to adverse soil temperature and moisture than preadults and adults (Sutherland and Sluggett, 1974). The adult of *Cynipanguina danthoniae* is the resistant stage, however, which is unusual for the *Anguina* group (Maggenti et al., 1973).

Flooding

Excessive moisture, from either irrigation or rainfall, can cause a reduction in nematode numbers or in the severity of the ensuing disease (Hollis and Fielding, 1958; Cralley, 1957; Van Gundy et al., 1968). Death is not always a direct effect of water per se but may occur by starvation, suffocation, or possibly by toxins such as has been suspected of *Clostridium* against *Tylenchorhynchus martini* (Johnston, 1957). In some instances, a reduction in nematode numbers results simply from dilution by dispersal as might happen with aquatic plants. Irrigation can disseminate plant-parasitic nematodes, however, and could conceivably magnify a problem rather than mitigating one (Faulkner and Bolander, 1970a, 1970b).

Conclusions have varied concerning flooding as a practical means of control. Watson (1921) believed that a week or two of flooding was usually necessary to reduce numbers of nematodes, if not control them. Brown (1933) found that juveniles of *Meloidogyne* were killed in land submerged for 4 months but that the eggs remained viable. No nematodes were found after 22½ months, and no infection occurred after the land was planted to a crop. Flooding was practical for nematode control in some instances, and the effects could be seen in the following year's crop (Thames and Stoner, 1953).

AERATION AND SURVIVAL

When an area is flooded, the oxygen available to nematodes decreases and carbon dioxide usually increases. Except in bodies of more or less permanent water, the oxygen supply may be reduced only temporarily, as it is during rainy periods. If the flooding period is short, kill may be small and the nematodes may merely go into a quiescent state. When oxygen was reduced from 20 to 5% (being replaced by carbon dioxide, with nitrogen maintained at 80%), initially the hatch rate of *Meloidogyne javanica* eggs decreased, but there was essentially no difference in total hatch after three weeks (Wallace, 1968a). No hatch occurred at 0% oxygen. Increasing the time that the eggs were in an oxygen-free atmosphere reduced subsequent hatch in aerated water. The data also indicated that absence of oxygen may be lethal to the embryonic stages. There was no change in activity of hatched juveniles kept in an oxygen-free environment for 4 days, but activity was greatly reduced when the juveniles were kept free of oxygen for 8 days and then placed into aerated water. It was concluded that hatch is inhibited by low diffusion of oxygen, with consequently small concentrations. It also was pointed out that even relatively little soil drainage can provide adequate aeration. Low aeration can kill embryos, but it also maintains infectivity in juveniles by inducing quiescence, thus perhaps extending the survival time (Van Gundy et al., 1967; Wallace, 1968a). Wallace (1968b) found that the egg sac can shrink and expand as soil moisture varies. Under low moisture conditions, shrinking inhibits hatch within the protection of the egg sac matrix. He suggested that damage by *M. javanica* may be larger in soils with a large crumb structure when there is plentiful irrigation or rainfall.

As is common for other phenomena, differences in developmental stages and species are variously affected by different gas tensions. Van Gundy and Stolzy (1963) found that molting and male survival in *Hemicycliophora arenaria* were the parts of the life cycle most sensitive to insufficient oxygen, while survival and movement of fourth-stage juveniles and of females were affected least. Van Gundy et al. (1962) also found that reduced oxygen tensions were unfavorable for hatching and reproduction of *Tylenchulus semipenetrans*. *Trichodorus christiei* and *Xiphinema americanum* were more sensitive to reduction of oxygen tensions than were *T. semipenetrans* and *Meloidogyne incognita*. Numbers of *Aphelenchus avenae* and *Caenorhabditis* were reduced significantly at 5% oxygen and inhibited at 4% oxygen and below in laboratory tests (Cooper et al., 1970). Aeration of 21% to 10% oxygen had no effect on reproduction. Five days or less of fluctuations between high and low oxygen concentrations inhibited buildup of *H. arenaria* on tomato in soil and of the other nematodes in vitro.

Hemicycliophora arenaria was inhibited by exposure to nitrogen for 12 hours every three days. As little as 4 hours exposure to nitrogen every three days inhibited *A. avenae* and *Caenorhabditis.*

EFFECT OF TEMPERATURE ON SURVIVAL

Temperature is also a major factor limiting distribution of plant-parasitic nematodes. As might be surmised, extremes may be inhibiting and survival is often best at a moderate temperature (Bergeson, 1959; Figure 8.2; Elmiligy, 1971).

It is evident that much could be learned of survival in cold temperatures by studying nematode distribution and soil temperatures. The use of Weather Bureau data, however, contains inherent risks due to the existence of microclimates. Soil temperatures often vary considerably within inches in the soil profile. Similar differences can occur within yards horizontally, especially where there is a change of soil type. Summaries of air temperature data as presented by the Weather Bureau denote general overall temperature changes and possibly reflect general relative changes in soil temperatures, but the soil microclimates are so diverse that care should be taken in using such data. Snow is a great insulator; in some winters frost lines may never form in the soil, owing to large accumulations of snow.

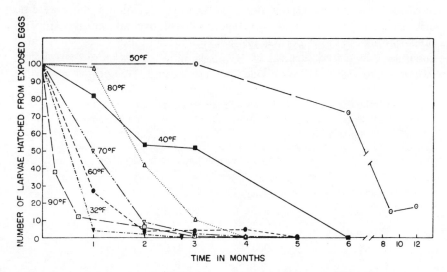

Figure 8.2. The influence of temperature on the survival of eggs of *Meloidogyne incognita acrita* in soil. Number of larvae determined by gall counts on an indicator plant. (From Bergeson, 1959, p. 348.)

Differential survival of nematodes based upon temperature is known, but direct evidence is meager. Certainly many tropical species are limited northward by cold temperatures. Even in the temperate regions, thermal differences are evident. *Meloidogyne incognita* and *M. incognita acrita* are the most common species of the genus south of the latitude of Washington, D.C. (Taylor and Buhrer, 1958) but are reported infrequently north of there. In support of this, Sayre (1963) found that *M. hapla,* a species common in the northern latitudes, could survive the winter at Harrow, Ontario, but that *M. incognita* and *M. javanica* could not. Similarly, *M. hapla* eggs withstood -2 C much better than those of *M. javanica*, and *M. hapla* eggs were less tolerant of 36 C than those of *M. javanica* (Daulton and Nusbaum, 1961). In addition, Bergeson (1959) reported that *M. hapla* juveniles and eggs could survive cold temperatures better than *M. incognita.* Adults of *Tylenchorhynchus martini* survived the winter better than juveniles in Minnesota (Patel and MacDonald, 1968), and Rhoades and Linford (1961b) found that preadults of *Paratylenchus projectus* could withstand sudden exposure to low temperatures better than other stages (See Table 8.2).

Preconditioning nematodes by exposing them to progressively lower temperatures or by alternating temperatures also affects their survival. Following earlier work in which he found that *M. hapla*, but not *M. incognita*, could survive winters in Ontario, Sayre (1964) found that the ice formed in eggs and juveniles during rapid cooling of 10 C per minute was lethal to both species. Under slower cooling at 1 C/minute, there were no differences in the ability of these species to survive subfreezing temperatures. Storage for long periods just above freezing, followed by subfreezing conditions, increased survival of *M. hapla* significantly more than that of *M. incognita*. Sayre (1964) divided nematodes into three groups according

Table 8.2. Survival of the stages of *Paratylenchus projectus* subjected to various temperatures for 96 hours. After Rhoades and Linford (1961b)

Temperature (C)	Nematodes extracted from 10 g of soil		
	Second- and third-stage larvae	Preadults	Females
Original soil	361	387	112
22	606	345	172
7	420	376	73
-7	4	349	5
-12	0	119	0
-19	0	42	0

to the injury resulting from progressively reduced temperature (Figure 8.3). One group included nematodes that are susceptible to chilling and are killed above freezing. Nematodes in the second group escape freezing at least temporarily by supercooling, and nematodes in the third group are not injured by freezing. Miller (1968) speculated that endoparasites such as *Pratylenchus penetrans* were more susceptible to freezing than the ectoparasites, who spend a good portion of their life in the soil. However, temperatures of −12, 0, and 4 C had little adverse effect on survival of *Ditylenchus dipsaci* juveniles in leaf tissue in soil (Miyagawa and Lear, 1970).

Soil conditions and duration of exposure also influence survival. Working with *P. penetrans* in New York, Kable and Mai (1968b) found that all motile stages overwintered in soil and senescent roots, but in the roots there were no significant differences in survival in different soil types or at different depths. Significantly more nematodes survived at 30 cm than at 15 cm in a sandy-loam soil, but this was reversed in a clay-loam. There were significantly more nematodes in the roots than in the soil at 15 cm in a sandy-loam, but there was no difference at 30 cm. In a clay-loam, survival was greater in soil than in roots at 15 cm, but at 30 cm depth survival was greater in roots than in soil. The reasons for this may be complex, and much study is needed. Kable and Mai found that in laboratory studies numbers of *P. penetrans* surviving in soil decreased with increasing moisture and temperature above freezing. None survived 15 days, however, at −15 C (Kable and Mai, 1968a). Storage of *Rotylenchulus parvus* in fallow soil for 7, 14, 21, and 28 days at 25, 15, 4, and 1 C resulted in a population decrease with the longer periods of exposure and colder temperatures (Dasgupta and Raski, 1968).

Although it has been shown by using liquid nitrogen that nematodes, especially free-living ones, can survive temperatures of −196 C (Hwang, 1970) or below, such temperatures never occur in the field, and so this information is of little practical use to us here. It may be developed into a good storage tool (Sayre and Hwang, 1975), however, and may eventually shed more light on the physiological processes involved in freezing of nematodes.

Although much emphasis has been placed on survival at low temperatures, it is evident that moderate and high temperatures also are important in survival (Table 8.3). This, of course, will vary with the nematode species and the duration and conditions of exposure. When nematode-infested soil was held at temperatures of 2, 13, 24, and 32 C, *Helicotylenchus dihystera* were recovered at 220 days at 2 and 13 C, at 80 days at 24 C, and no longer than between 65 and 80 days at 32 C, indicating that this species can tolerate cooler temperatures better than warmer ones. *Tylenchorhynchus claytoni*, however, was recovered only for the first 40 days, and survival was somewhat better at the warmer than at the cooler temperatures. *Paratylen-*

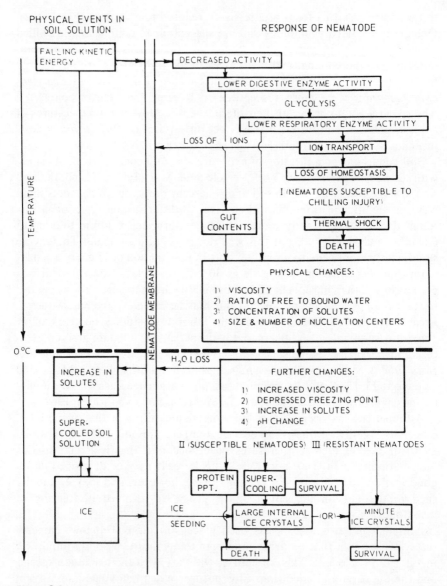

Figure 8.3. A schematic diagram to illustrate how groups of nematodes in soil solution might react to decreasing temperatures. (From Sayre, 1964, p. 177.)

Table 8.3. Upper lethal temperatures for survival of nematodes

Nematode	Temperature (C)	Duration	Condition	Reference
Helicotylenchus dihystera	32	65–80 days	Air dry soil	McGlohon et al., 1961
Heterodera schachtii	52		Survived 3 months summer fallow at 52 C	Thomason and Fife, 1962
H. tabacum	32 and 37	6 days	In field capacity soil; killed emerging juveniles but not eggs	Miller, 1969
H. tabacum	42	24 hours	Encysted eggs in soil	Miller, 1968
Hoplolaimus galeatus	32	0–2 days	Air dry soil	McGlohon et al., 1961
Meloidogyne hapla	45.5	1 hour	Dormant rose stock	Martin, 1968a
M. javanica	46–47.5	2 hours	In potato tuber	Martin, 1968a
Paratylenchus projectus	32	0–2 days	Air dry soil	McGlohon et al., 1961
Trichodorus christiei	32	0–2 days	Air dry soil	McGlohon et al., 1961
Tylenchorhynchus claytoni	32	40–50 days	Air dry soil	McGlohon et al., 1961

165

chus projectus, Hoplolaimus galeatus, and *Trichodorus christiei* were not recovered at any temperature after two days of air drying (McGlohon et al., 1961). When infected potato plants were exposed to temperatures above optimum for *Heterodera rostochiensis* (up to 38 C), the number of juveniles that became adults decreased with increasing temperature. The most sensitive stage was the preadult male (Trudgill, 1970).

AGING AND SURVIVAL

It is not surprising that a decrease in infectivity occurs with a decrease in motility and food reserves of nematodes, as Van Gundy et al. (1967) have shown for *Meloidogyne javanica* and *Tylenchulus semipenetrans.* Survival of infective juveniles was greater in soil than in water and was associated with greater retention of food reserves. The principal storage material in the intestine was lipid. The length of time that the nematodes were motile and infective was 128 days for *Tylenchulus semipenetrans* but only 32 days for *M. javanica*, which had a higher respiration rate than *T. semipenetrans.* The food reserves were used up most rapidly at high temperatures, in dry soils, and in oxygenated solutions, all these conditions resulting in decreased motility and infectivity.

SURVIVAL IN ANIMALS

The survival of nematodes through the alimentary canal of larger animals has been known or suspected for some time. Survival and dissemination are separate phenomena, but if a nematode can survive passage through the alimentary tract of animals, dissemination is a matter of movement of the carrier during the time it contains the nematodes. After the nematode passes from the digestive tract of the vector, its survival is governed by the external conditions including the deposited excrement. When 8-day-old sheep were fed beets containing cysts of *Heterodera,* presumably *H. schachtii,* uninjured females of this nematode were found in the excrement, which was believed to be the source of some infections (Girard, 1887). Other early workers were less successful, however. Davaine (1857) concluded that *Anguina (Tylenchus) tritici* juveniles were digested by the chicken, dove, and sparrow but passed unchanged through the digestive tract of the frog, salamander, and goldfish. Marcinowski (1910) found that when several animals, including birds, rodents, and sheep, were fed galls of *A. tritici*, the excrement contained at least a few living nematodes. More recently, it has also been shown that plant-parasitic nematodes can pass through other animals and still be infective to plants. Martin (1968b) found that if cows were fed potato tubers infested with *Meloidogyne arenaria,* small numbers of eggs or juveniles could pass through the digestive tract and remain infective. Similar results were obtained using *M. javanica* and the common mole rat (Martin, 1969). Since the mole rat lives much of the time under the ground, and since much of its diet consists of roots, it is highly probable that it disseminates some nematodes.

As with *Meloidogyne* spp., not all nematodes of *Heterodera* that pass through the digestive system are infective. Of 18,000 *H. glycines* cysts ingested by pigs, 19.4% were recovered from the feces. Six juveniles emerged from the cysts but were highly vacuolated and were not infective (Smart and Thomas, 1969). Positive results are generally more meaningful than negative ones, however, and it is probable that other animals can transport nematodes through their digestive systems.

It has been speculated for years that birds might transport nematodes either externally or internally, especially in conjunction with new infestations of *Heterodera glycines*. Epps (1971) found that when birds were force fed eggs and juveniles of *Heterodera glycines*, infective juveniles were recovered in the excrement. Infective juveniles also were recovered in the excrement of birds fed cysts 24 to 48 hours previously. Cysts were recovered from the digestive systems of starlings captured in *H. glycines* infested fields. Even though the recovery was small, it indicates that the nematode probably is disseminated by birds and that new infestations theoretically can start from nematodes transported this way.

Brodie (1976a), on the other hand, believed that migratory birds are unlikely disseminators of *H. rostochiensis*. Viable juveniles emerged from cysts that passed through the birds' digestive tract within one-half hour. Juveniles were not viable that remained in the digestive tract for more than 1½ hours for starlings, more than 1 hour for pigeons, or more than ½ hour for four other species of birds. Cysts that remained in contact with the excrement for more than 72 to 96 hours after passage failed to hatch regardless of the time within the digestive tract. Since both survival and dissemination are involved, such work indicates that dissemination of viable cysts is possible. Excrement in nature is subject to diverse environmental conditions; some of these may shorten nematode survival time, but it may be equally true that some conditions may lengthen nematode survival time.

CHAPTER 9

PARASITISM OF NEMATODES IN BIOLOGICAL CONTROL

In Chapters 7 and 8 we discussed the physical and chemical factors that either restrict the population growth of nematodes or limit their survival. Biological factors are also important, and we will discuss some of them in this chapter.

Sewell (1965) defined biological control as "the induced or natural, direct or indirect limitation of a harmful organism, or its effects, by another organism or group of organisms." This broad definition has its usefulness, although "harmful" could be deleted. Under such a definition, one must include antagonism, host resistance, and tolerance. Although resistance is a form of biological control, it is usually treated separately. Here we shall deal entirely with biological control by parasitism and predation.

Parasitism and predation occur throughout nature. Thus there is reason to assume that there is natural biological control of nematodes wherever life processes are even moderately sustained. Describing the extent to which control occurs, the mechanisms involved, and the possibilities of altering nature to the benefit of humans is a difficult task. In the latter, we should consider the long-term results and broad-based environmental consequences. Generally the soil is well buffered, and results that occur in the laboratory might be difficult to duplicate in the field.

Early optimism for biological control of nematodes has changed somewhat to pessimism. Nevertheless, along with increased emphasis on our total environment and with the probably warranted restrictions on future use of some agricultural chemicals, every effort should be made to take advantage of biological control measures. As Baker and Cook (1974) have pointed out, difficulties and failures in biological control have probably been overemphasized. Empirical studies should not be judged or dismissed too quickly, until predator–prey relationships and other aspects of biological control are studied in toto. Extraction methods that depend on nematode mobility, such as the Baermann funnel, will not recover dead or sluggish nematodes (Williams, 1967). Thus the occurrence of parasites and their probable importance is greater than our present knowledge might indicate.

Predator–prey relationships are an important part of energy flow in the biological world, and if some relationships are not obvious it does not follow that they are not meaningful. There is much to learn of predator–prey relationships with nematodes and of the possibilities of altering them to our benefit. It is possible that man hinders rather than helps biological control. For example, nematicides kill predacious nematodes as well as parasitic ones. Does this allow the surviving plant parasites to increase faster than they would if chemicals had not been applied? Even though there has been little practical increased biological control of nematodes through direct human influence, let us examine some of the successes, failures, possibilities, and problems that have occurred.

BACTERIA DETRIMENTAL TO NEMATODES

It seems logical to expect that bacteria can infect nematodes and deplete their populations. Few reports of bacteria infecting plant-parasitic nematodes exist, although there are many records of bacteria occurring in free-living forms (Dollfus, 1946). In one instance, *Xiphinema americanum* collected in West Virginia was infected with *Pseudomonas denitrificans* (Adams and Eichenmuller, 1963). The bacteria were distributed throughout the body but were sparse in the esophagus of juveniles. In adult females, the bacteria were concentrated in the intestine and ovaries, which suggests possible transovarial transmission. Infection apparently did not affect nematode movement, however. In such instances the nematode population may not be reduced unless infection of the ovary reduces egg production.

Other bacteria are detrimental without actually invading the nematode. The bacterial feeders *Caenorhabditis briggsae, Rhabditis oxycerca,* and *Panagrellus* sp. were lysed by myxobacteria in agar and liquid media. Disintegration of the nematodes was marked within 6 hours (Katznelson et al., 1964). *Aphelenchus avenae* and juveniles of *Heterodera trifolii* were not affected. Johnston (1957) reported that a *Clostridium* species produced a toxin that was lethal to nematodes within 2 minutes. He concluded that anaerobes may be a factor in the low survival of nematodes in water-saturated soils. Later, Hollis and Rodriquez-Kabana (1966) found and inverse correlation with the combined values of butyric and propionic acids, which are produced by *Clostridium butyricum*, in flooded cornmeal-amended soil. This phenomenon was not attributed to a pH effect.

NEMATOPHAGOUS FUNGI

Some of the most intriguing associations of nematodes with other organisms are those of nematode-trapping and other parasitic fungi. Although Soprunov (1958) argues in favor of "predacious" as the term for trapping forms, it seems that this expands the common usage of the term with possible teleological implications. Whatever term one applies, the fact remains that these fascinating fungi have received much attention as possible agents for biological control. The major pioneer works in the field have been the many descriptions of these fungi by Drechsler, the review of all work until 1946 by Dollfus (1946), and the thoroughly enjoyable if somewhat optimistic book by Duddington (1957). The methods by which these fungi capture and consume nematodes have excited the imagination, and the reader is referred to the works cited, among others, for descriptions and life habits.

It is a curious fact that these fungi have seldom been included in lists of soil fungi, and yet many species are more common than many better known fungi. Nematode-parasitic fungi can be isolated readily by rather special but simple techniques. The trapping mechanisms are known to occur in soil; the fungi have been found attached to living and dead nematodes freshly washed from soil samples (Linford and Yap, 1939). As cultured in the laboratory, many of these fungi behave differently in the presence of nematodes than otherwise. Endoparasitic fungi often are not recovered unless nematodes are present.

As discussed in the section on organic matter amendments in Chapter 7, there are many instances in which different forms of organic matter added to soil have been beneficial by either increasing yields or reducing disease severity. This has led to speculation that part of the nematode control could be due to an increase in nematode-parasitic fungi. An early experiment that attracted attention involved large portions of pineapple plants that were incorporated in root-knot-infested soil planted with cowpea (Linford et al., 1938). A marked reduction of galling was evident at harvest where chopped pineapple was added, compared with the control treatments. Comparable results were obtained when para-grass or cane sugar was used as organic matter. After decomposition of the organic matter, many kinds of nematodes, including predators, often increased. It was speculated that these and other organisms might be reducing the root-knot nematode population directly or indirectly. Many attempts to infest the soil with predacious and parasitic fungi followed in the hopes that the fungi would significantly reduce the severity of disease and plant-parasitic nematode populations. Linford and Yap (1939) added five species of nematode-trapping fungi to soil infested with *Meloidogyne* sp. and found that plants in treatments with *Dactylella ellipsospora* had significantly less nematode injury after 15 months than those in treatments where fungi were not added.

It is frequently difficult to ascertain whether or not nematode populations are being decreased by the treatments, even though the disease severity may be lessened. Seven of eight organic compounds, however, reduced the number of *Tylenchulus semipenetrans* in citrus (Mankau and Minteer, 1962). Hams and Wilkin (1961), using pot tests, reported some success in reducing invasion of plants by *Heterodera* but had little success in the field when trapping fungi were introduced with organic matter of various kinds. They likewise had no success when the cyst-invading fungus *Phialophora heteroderae* was used. These authors recognized that the ecology of the fungi may be important in their establishment.

Other attempts have been made to control *Heterodera* spp. with addition of organic matter and fungi, but the results have not been spectacular. One-half kilogram of *Dactylaria thaumasia* mycelium was added per 0.000067 ha plot, with and without bran, in attempts to control *H. schachtii* (Duddington et al., 1956). The number of cysts decreased, but not the final egg population. Crop yields increased about equally whether bran or only fungus was added. The authors were not sure if the fungus was controlling the nematode or if it was acting as just another organic matter additive. Addition of large quantities of mycelium to soil would pose major engineering problems. In later work, the number of *Heterodera avenae* in roots was reduced significantly when a green manure, or *D. thaumasia*, was incorporated in microplots (Duddington and Duthoit, 1960; Duddington et al., 1961). In spite of these results, other investigators have been less successful. Negative results generally were obtained in observations and tests with the following combinations: *Dactylaria eudermata* and *Heterodera rostochiensis* (Hutchinson and Mai, 1954); *Arthrobotrys musiformis* and *Radopholus similis* (Tarjan, 1961); *Dactylaria thaumasia, Arthrobotrys arthrobotryoides*, and *Meloidogyne incognita* (Mankau, 1961a); trapping fungi and *M. incognita* on tomato and okra (Mankau, 1961b); and trapping fungi and *Hoplolaimus tylenchiformis* and *Pratylenchus pratensis* in barley (Hutchinson et al., 1960). There is evidence that the endoparasitic fungus *Catenaria anguillulae* is no more than a weak parasite because the greatest invasion occurred in nematodes killed by heat. Invasion was the least in nematodes not treated with heat and only slightly more in those weakened by heat (Boosalis and Mankau, 1965). Sayre and Keeley (1969) concluded that the isolate of *C. anguillulae* used by them was not a good control agent against *Panagrellus redivivus* and *Ditylenchus dipsaci*. On the other hand,

Esser and Ridings (1974) found that *C. anguillulae* attacked live nematodes more readily than dead ones. They also found the infectiveness varied with the isolate of the fungus.

Some of the inconsistent results noted may be partly a matter of technique, but it seems evident that the problems of making biological control of nematodes work on a field scale are formidable. It is recognized that perhaps a sounder approach is to investigate the basic ecology of the parasitic fungi and treat them as soil inhabitats. These fungi have their ecological niches, and they are subject to the elements as much as any other soil organism. When cabbage leaves or sugars were added to naturally infested soil and the nematophagous fungi observed for 11 weeks, the retiary trap-forming kinds, of which *Arthrobotrys oligospora, Dactylaria psychrophila,* and *Trichothecium cystosporium* were present, peaked and declined in the first half of the test period. *Dactylella cionopaga* became dominant during the second half of the period and then declined. A ring-forming fungus was observed only in the final stage of succession. Endoparasitic fungi, including *Acrostalagmus obovatus, Harposporium anguillulae,* and *Nematoctonus leptosporus,* occurred sporadically. These successional patterns appeared to bear no relation to the presence of nematodes, however (Cooke, 1963). This supported previous work in which it was frequently found that nematode-trapping fungal activity decreased even though nematode populations increased (Cooke, 1962a). This indicates that the presence of nematodes is not the sole determining factor in the ability of fungi to trap nematodes. It has been found, however, that there was greater growth of fungi in agar culture when nematodes were present than when they were absent (Monoson, 1968).

There has been interest in the mechanism of trap formation ever since Comandon and de Fonbrune (1938) reported that a sterile culture filtrate in which nematodes had lived previously could stimulate trap formation. There is evidence that a toxin is produced by *Arthrobotrys oligospora* (Olthof and Estey, 1963). Pramer and Stoll (1959) reported that a water-soluble substance, which they named "nemin," is a metabolic product of the nematode *Neoaplectana glaseri* and was responsible for trap formation in *Arthrobotrys conoides.* Traps were not formed if nematodes were not previously in the test culture filtrate. Trapping organs of *Dactylella bembicodes, D. cionopaga, D. ellipsospora,* and *D. drechsleri* formed differentially to nemin produced by *Panagrellus redivivus,* indicating that specificities exist (Feder et al., 1963). Quantitative differences in the ability of five different nematode species to induce traps were noted by Monoson et al. (1974). That chemical composition of the substrate is important in trap formation was shown by Miura (1966), who found that *Arthrobotrys dactyloides* formed more traps spontaneously when grown on sodium nitrate, hippuric acid, or glycine-glucose media than on other media. Evidently a rich food base is necessary to sustain the nematophagous activity in nature. When sucrose was added to nonsterile soil, the population of free-living nematodes and the activity of trapping fungi both increased. But at a certain point in sucrose decomposition, trapping activity of the fungi ceased. Addition of more sucrose resulted in an increase in nematodes and a decrease in the predacious activity of the fungi. Cooke (1962b), using nematode-free soil, reported that it seemed that nematodes are necessary to initiate trapping organs but that an organic source other than nematodes is necessary to sustain the predacious habit. Olthof and Estey (1966) also found that *A. oligospora* needs a nematode-free organic source. Nutritional studies with *Arthrobotrys conoides* and *A. dactyloides* indicate that these fungi are not unique in their requirements (Coscarelli and Pramer, 1962; Miura, 1966). Biotin, thiamine, and zinc are required for maximum growth of *A. conoides*, a situation not unusual with other soil fungi (Grant et al., 1962). The greatest measurable trapping habit of *Arthrobotrys robusta*, however, was on media deficient in biotin or with nitrite as a nitrogen source (Hays and Blackburn, 1966). As expected, temperature affects the growth and trapping ability of nematophagous fungi (Monoson, 1968; Olthof and Estey, 1965; Soprunov, 1958).

Although pH readings were not recorded, Couch (1937) added varying amounts of H_3PO_4 and found that traps of *Dactylella bembicodes* occurred on an acid medium. *Arthrobotrys dactyloides* produced the most mycelium at pH 6.5 and lesser amounts toward the extremes tested of 4.5 and 9.5 (Miura, 1966). Soprunov (1958) reported that *Arthrobotrys conoides, A. longispora, A. oligospora,* and *Trichothecium globosporum* did not grow at pH 2.5 and growth was slight at 9.5. Generally excellent growth was obtained between pH 5.0 and 8.0. Thus there seems to be nothing unique about the general pH requirements for growth until extreme ranges are attained. Sayre and Keeley (1969), however, found that more infection of *Panagrellus redivivus* and *Ditylenchus dipsaci* by *Catenaria anguillulae* occurred at pH 8.0 or 9.0 than at pH 5.0, 6.0, or 7.0.

There is also evidence that nematode-trapping fungi are subject to the same antagonisms as other fungi. *Trichoderma* and *Gliocladium* have been shown to be antagonistic to *Arthrobotrys musiformis* on agar plates (Tarjan, 1961), and similar inhibitory phenomena doubtless occur in the soil. Although there was little fungistatic effect on germinating spores of several nematode-trapping Hyphomycetes (Cooke, 1964), conidia were inhibited by an agent in the soil (Cooke, 1968). It is uncertain how much of this is nutritional, but four constricting ring fungi generally did not have the ability to utilize nitrites and nitrates. Fungi that produced adhesive networks generally could utilize these substances. It is possible that the constricting ring fungi obtained their nitrogen from the host. Mankau (1962) found a water-diffusible factor in soil that is fungistatic to conidia of *Arthrobotrys arthrobotyroides, A. dactyloides,* and *Dactylella ellipsospora.*

While evidence is fragmentary, there seems to be nothing unique about the nematode-parasitizing fungi compared with many other soil fungi. Fungi that parasitize nematodes are ubiquitous in the soil, and doubtless much parasitism occurs naturally. The problem of selectively increasing the nematode-parasitizing fungi is a management problem on which there has been little progress. This, however, should not deter further work on these interesting fungi since their biology is so little known. It is doubtful that they can be put to practical use until the more fundamental questions are answered.

ANIMALS AS PARASITES OF NEMATODES

Amoebae Feeding on Nematodes

Amoebae are known to attack nematodes in the laboratory, but their importance in nematode control in natural ecosystems has yet to be evaluated. *Theratromyxa weberi* was seen engulfing nematodes in 20 minutes to 2 hours (Weber et al., 1952). The amoeba was a poor control of *Meloidogyne incognita* in greenhouse tests, however (Sayre, 1973). The nematodes that were attacked did not exceed 1 mm and included *Heterodera rostochiensis, Meloidogyne* juveniles, *Ditylenchus dipsaci, Pratylenchus pratensis,* and *Hemicycliophora.* Van der Laan (1954) found that large nematodes such as *Mononchus* and *Dorylaimus* did not serve as prey. He also recorded that 128 *Heterodera* juveniles were engulfed in one digestive cyst of *T. weberi.* Eight thousand individuals of *Pratylenchus* were consumed by 400 amoebae within 5 days. Impressive as this is under laboratory conditions, there was little propagation of the amoeba in pot tests. It was questioned that there would be effective biological control, since amoebae cannot survive much desiccation, their movement is slow, and evidently their prey is nonspecific. Winslow and Williams (1957) also questioned the effectiveness of amoebae in biological control.

A *Urostyla* sp. has been observed to ingest nematodes under certain circumstances but was apparently unable to digest them. In some instances, the nematodes were able to escape unharmed; in others, the amoebae, with nematodes in them, ruptured and disintegrated (Doncaster and Hooper, 1961). Esser and Sobers (1964) also reported that nematodes have been released or have escaped from amoebae.

Biological control of nematodes by amoebae offers little hope at present. It should not be expected that amoebae or any other organism will be highly effective under all conditions. Soil habitats are varied, and it is probable that a system that works in one ecological niche will be ineffective in another.

Sporozoan Parasites

Sporozoans, which comprise the nonflagellate protozoa, have been reported on or in nematodes on many occasions. Dollfus (1946) summarized studies prior to 1946 and cautioned on the probability of many errors and on the uncertain relationships of the organisms described. This was reiterated by Williams (1960). Much early work was by Micoletzky (1925), and most of the nematodes parasitized were nonstylet forms, such as *Trilobus*, and *Dorylaimus*, of which *D. carteri* (= *Eudorylaimus carteri*) was mentioned frequently.

Canning (1973) discussed protozoal parasites and mentioned that some of the large stomal nematodes, such as some of the free-living predators and bacteriophagous nematodes, could ingest oocysts of protozoa, but that the stylet canal of plant-parasitic nematodes is too small to serve as a port of entry for the cysts. It is unlikely, therefore, that these forms of protozoa are candidates for successful biological control of plant-parasitic nematodes.

A species perhaps recognized as having the best potential for biological control is *Duboscquia penetrans*, an organism whose taxonomy is controversial (Birchfield and Antonopoulos, 1976; Mankau, 1975). Thorne (1940) noted that 66% of 131 individuals of *Pratylenchus pratensis* from soil around cotton were infested with *D. penetrans*. Williams (1960) recorded that at least 34% of 174 females of *Meloidogyne* contained an organism that might be *D. penetrans*. In greenhouse tests there was a 50% reduction in an infected population of *Pratylenchus scribneri* within 55 days, but large populations of *Meloidogyne javanica* could be maintained in infested soil (Prasad and Mankau, 1969). Prasad and Mankau found that spores could survive and remain infective when stored in air-dried soil for 6 months. Further evidence of host specificity was found when *Meloidogyne javanica, M. arenaria,* and *M. incognita* became more heavily infected than *M. hapla* or *Pratylenchus scribneri* under similar conditions (Mankau and Prasad, 1977). Other species, including *P. brachyurus, Aphelenchus avenae, Ditylenchus dipsaci, Heterodera schachtii, Trichodorus christiei, Tylenchorhynchus claytoni,* and *Xiphinema index* did not become infected.

Tardigrades as Predators

There are several reports that tardigrades devour nematodes (Esser, 1963; Esser and Sobers, 1964; Hutchinson and Streu, 1960). Linford and Oliveira

(1938) found them in root and soil samples and mentioned their possible potential as control agents. Nothing is known about their effectiveness in nature. Tardigrades are usually polyphagous, and nematodes comprise only a part of their diet. Thus, if other forms of food are readily available, nematode populations may not be reduced drastically unless there is considerable selectivity. It is probable, however, that in certain microhabitats, nematodes serve as a partial food source for tardigrades.

Observation indicate that tardigrades feed on nematodes nonselectively. Hutchinson and Streu (1960) found tardigrades feeding on *Trichodorus aequalis* and *Tylenchus* sp. Doncaster and Hooper (1961) stated that they showed no preference for size of nematode prey and speculated that a toxin might be involved, because nematodes became immobile even after the tardigrade (*Macrobiotus* sp.) detached itself. *Hypsibius myrops* significantly reduced populations of *Meloidogyne incognita, Ditylenchus dipsaci*, and *Panagrellus redivivus* when the tardigrade fed on them for 3 days (Sayre, 1969).

Mites Feeding on Nematodes

Mites have been seen feeding on *Meloidogyne* (Linford and Oliveira, 1938) and *Heterodera* cysts (Murphy and Doncaster, 1957), but there is little information on the efficiency of mites in reducing nematode populations. While studying mite populations in manure piles, Rodriguez et al. (1962) found that three species of nematodes were prey of the mite *Macrocheles muscaedomesticae*. *Rhabditella leptura* proved to be a better substrate than *Panagrolaimus* or *Diplogaster*, and the investigators were able to raise the mite to maturity on *R. leptura*. Obviously, much must be accomplished before we can understand the importance of these associations.

Enchytraeids as Parasites of Nematodes

Schaerffenberg and Tendl (1951) reported control of *Heterodera schachtii* by enchytraeids invading the root tissue and predigesting the nematode juveniles. Boosalis and Mankau (1965) questioned this because enchytraeids are normally saprophagous.

Predacious Turbellaria

Although turbellarians have been known to use nematodes as food, there is little critical work on the subject. Sayre and Powers (1966) found an undescribed species that could ingest nematodes, but the number of eggs produced by the turbellarian differed with the nematode prey. When a relatively constant biomass of three nematode species was maintained in a water medium, the turbellarian laid approximately 10 eggs in 9 days when feeding on *Panagrellus redivivus*. The turbellaria laid only an average of one egg each when fed *Meloidogyne incognita* juveniles, and only two when they were fed *Ditylenchus myceliophagus*. Eggs produced from turbellaria fed *P. redivivus* began hatching on the ninth day, and the numbers increased rapidly. Hatching of eggs from turbellaria fed with the other two nematodes was poor. These selective differences could not be attributed solely to the larger size of *P. redivivus*, but it probably had some effect. With three particle sizes of sand and *M. incognita*, the number of surviving juveniles was greatest in the smallest (421 to 595 μm) particles. This indicates that chance encounters between the predator and prey are decreased in the particulate soil. Fewer galls occurred on tomatoes when the seedlings were planted at the same time that the turbellaria and root-knot juveniles were added to the soil than when treatments contained only nematodes. Galling was significantly reduced, however, when the flatworms and nematodes were introduced into the soil 48 hours before the tomato plants. This seems to indicate that the turbellaria consumed the nematodes before the juveniles had time to penetrate the roots. Although the authors demonstrated that turbellaria had an adverse effect on nematodes, they doubted that the flatworms with which they worked could give economic control although they possibly could be used along with other forms of control. This conclusion was based partly on comparison of populations of turbellaria recovered from the soil with those necessary to give partial control. The polyphagous nature of turbellaria probably would reduce further their effectiveness for nematodes.

Predacious Nematodes

For about 20 years beginning in 1917, Cobb, Christie, Thorne, and Linford were active in cataloging and describing the habits of nematodes that feed on other nematodes or other small soil organisms. Activity declined until the 1950s and 1960s, when interest was renewed. At present, however, relatively little is known about which nematodes prey on other nematodes and whether or not they are effective in reducing nematode populations in the field. As Christie (1960) pointed out, this is an attractive and unexplored field.

Many genera of nematodes are known to prey on other nematodes. Some of the best known are *Sectonema, Nygolaimus, Mononchus, Mylonchulus,* and *Seinura*. Cobb (1929) listed 16 genera of free-living nematodes, many of them aquatic, that contain carnivorous species. Some are known to feed on animals other than nematodes. Predacious nematodes are sometimes divided into three groups (Christie, 1960) according to their feeding habits. Those of the first type, such as *Tripyla* have a rather plain esophagus and swallow other nematodes whole. They may also feed on other small animals such as protozoa and rotifers. In the second type, the predators either swallow other nematodes whole or puncture the larger ones and suck out their internal contents. This type often has a large stoma armed with a single large tooth, often accompanied by several rows of smaller denticles. These include the well-known *Mononchus* and related genera. The third type includes stylet nematodes that puncture other nematodes with a protrusible odontostyle or stomatostyle. These include such distantly related forms as *Seinura* in the Secernentea and *Dorylaimus* in the Adenophorea.

It should be remembered that the primary diet for some suspected predators of nematodes may be mainly or entirely organisms other than nematodes, with nematodes being only incidental. In other instances, nematodes may be the primary food source. Thorne (1930) found that although *Nygolaimus* spp. and *Sectonema* were closely associated with maturing females of *Heterodera schachtii* clinging to the roots of sugar beets in the field, in the laboratory they fed on oligochaetes and not on nematodes. *Sectonema ventralis* was much more voracious than *Nygolaimus*. Both Thorne (1927) and Steiner and Heinly (1922) reported that *Mononchus papillatus* was very voracious on other nematodes, but Thorne also reported that *Mononchus macrostoma, Mylonchulus sigmaturus,* and *M. parabrachyurus* feed on protozoans, rotifers, and other microscopic soil organisms more than on nematodes. Linford (1937) noted that a *Dorylaimus* sucked nematode eggs and preyed on rotifers and large infusoria. Hollis (1957) found that *D. ettersbergensis* fed not only on nematodes, but also on infusoria, conidia of a fungus, and algae. A *Thornia* sp. has been observed to feed on *Tylenchulus semipenetrans, A. avenae,* other nematodes, encysted amoebae, fungi, and yeast cells (Boosalis and Mankau, 1965). Several species of *Seinura* feed or have been reared on one or more species of nematodes, the hosts being *Aphelenchus avenae, Aphelenchoides parietinus, Pratylenchus* sp., juveniles of *Meloidogyne hapla,* and *Ditylenchus dipsaci* and *Heterodera trifolii* (Hechler, 1963; Linford, 1937; Hechler and Taylor, 1966). That there is a host preference is indicated in that *S. tenuicaudatus* did not penetrate the cuticle of an adult *Xiphinema* sp. or *Hoplolaimus galeatus,* and there was no attempt to penetrate the white females or cysts of *H. trifolii* (Hechler, 1963). *S. tenuicaudatus* has also been observed to feed on eggs (Linford, 1937).

Relatively little is known about the ecology of predacious nematodes. Many members of the *Dorylaimoidea* are predacious and are common and widely distributed in diverse habitats (Thorne, 1939; Thorne and Swanger, 1936). Many nematodes are suspected of being predacious because of their morphology, but their food habits actually are unknown. The Mononchidae are widely distributed and their habitats vary. The species that Thorne (1927) studied in Utah appeared to thrive better in red sandy soils and were less common in heavy soils. Body size may be a factor for this, but with our limited knowledge, generalizations are tenuous. By noting the incidence of gravid females relative to soil temperatures, Thorne concluded that *Mylonchulus sigmaturus* reproduced at a lower temperature than *M. parabrachyurus*. The greatest reproduction of *M. parabrachyurus* occurred in April and May, and that of *M. sigmaturus* in March and April. *Mylonchulus parabrachyurus* migrated deeper into the soil during the hot summer months, but this was not true of other mononchs (Thorne, 1927). Since there was probably only one generation a year, the food habits were largely unknown, and populations of the mononchs were unstable, Thorne did not believe that predators were effectively controlling the sugar beet nematode.

Members of *Seinura* are widely distributed, but little is known about their natural populations. Some, such as *S. celeris, S. oliveirae, S. oxura,* and *S. tenuicaudata* have been found in cultivated fields (Christie, 1933; Hechler and Taylor, 1965; Mai et al., 1960), and *S. oahuensis* has been reported from a native Kansas prairie (Orr and Dickerson, 1966).

Little also is known of how predacious nematodes affect natural populations of other nematodes. No chemical attraction between the prey and host is known. Cobb (1920) states that the mononchs never follow other nematodes into plant roots except in open root cavities. The prey of *Dorylaimus* (Linford and Oliveira, 1937) and *S. tenuicaudatus* (Hechler, 1963; Linford, 1937) is found only by chance; if no contact is made, there is no pursuit by the nematode. Linford (1937) did not find cannibalism with *S. tenuicaudatus*, while Hechler (1963) reported that this species will become cannibalistic when the favored host is depleted.

Most studies involving spear-bearing nematodes have been with agar cultures in which a suitable host such as *Aphelenchus avenae* is propagated on a fungus. The predacious nematode then is introduced and the reactions noted. Under favorable conditions, the life cycle of many species of *Seinura* is rapid, being completed in 2 to 4 days for *S. celeris, S. oliveirae,* and *S. oxura* (Hechler and Taylor, 1966) and 5 to 7 days for *S. steineri* (Hechler and Taylor, 1966) and *S. tenuicaudatus* (Hechler, 1963). This rapid increase can reduce large numbers of the host in a short time. In spite of this rapid reproduction in the laboratory, it probably does not occur as fast in the soil, since *Seinura* is not very common in spite of a frequent abundance of *A. avenae*. Steiner and Heinly (1922) reported that individuals of *Monochus papillatus* produced from 21 to 41 eggs in the laboratory. As noted earlier, they also reported that this species was an effective predator. One individual killed 83 *Meloidogyne* juveniles in one day, and during a 12-week period one mononch killed 1332 nematodes. As Christie (1960) has pointed out, the reports of effectiveness of predacious nematodes in biological control have often been opinions based upon observations not supported by controlled experiments. Until suitable experiments are carried out under conditions closer to the natural environment than petri dishes in the laboratory, we will continue to be largely ignorant of the possibilities of any useful control. Boosalis and Mankau (1965) have come closer to the problem with experiments using potted citrus seedlings. One group of seedlings was infested with *Thornia* sp. plus *Tylenchulus semipenetrans*, while the other groups were infested with only *Thornia*, *T. semipenetrans*, or neither. There were no significant differences in fresh top weights of the seedlings after 17 months, and the differences in the citrus-nematode populations were not significant in pots to which they were added. *Thornia* sp. was numerous when *T. semipenetrans* was added but was nearly absent when there were no citrus nematodes. After 29 months, the citrus-nematode population was lower when *Thornia* was present, but the difference was not significant. The populations remained small in the *Thornia* treatments compared to the *Thornia* plus *T. semipenetrans*. The results were not striking, and much more must be done before generalizations can be justified.

Insects Feeding on Nematodes

There have been scattered observations of insects feeding on nematodes but few well-documented studies. Food balls that some ants make for their larvae were found to contain fungi as well as eggs and remains of nematodes (Cobb, 1924). Esser (1963) observed one dipterous larva devouring a sting nematode, *Belonolaimus*, and Brown (1954) reported that an insect that probably was an *Isotoma*, a spring-tail, took only 2 to 3 seconds to ingest a nematode. While working out methods for studying the soil meiofauna, Murphy and Doncaster (1957) reported that the Collembola *Onychiurus armatus* was the most voracious of those tested on *Heterodera cruciferae*. White females were preferred to brown cysts. The insect also was seen to feed on *Dorylaimus* and *Mononchus*. There was also some damage to females of *Heterodera schachtii* and *H. trifolii*. In a preliminary observations, about 7% of females or cysts of *H. cruciferae* had characteristic damage caused by *O. armatus*. Other Collembola seen to damage *H. cruciferae* to varying degrees were *Hopogastrura* sp., *Isotoma veridis*, and *Orchesella villosa*.

While these reports are interesting, we know practically nothing of the amount of natural insect predation that occurs or of ways to increase it.

NEMATODE VIRUSES _____

The only published report of nematodes being attacked by viruses is that of Loewenberg et al. (1959). These authors found that juveniles of *Meloidogyne incognita incognita* became sluggish in their movements and galls were not subsequently produced. After eliminating bacteria and fungi as possible causes, they found that filtrates of a suspension containing the sluggish nematodes contained an infectious agent. There were careful controls and continued checks to ensure that bacteria did not pass through the Seitz filter. The authors concluded that the filterable agent was probably a virus.

This is probably a more common phenomenon than is on record, but nothing is known on how much nematode populations are reduced by viruses. Extension of this work should prove fruitful.

CHAPTER 10

BIOLOGICAL INTERACTIONS

The biological constituents are as much a part of the environment as are the physical and chemical constituents.

Plant-parasitic nematodes often are considered as separate entities solely responsible for symptoms in an infected plant. Evidence indicates, however, that nematodes often facilitate entry into plants by other agents such as fungi, bacteria, or viruses. Association of nematodes with other organisms may contribute actively to the disease syndrome. Such associations make it difficult strictly to apply Koch's postulates. Fawcett (1931) warned long ago that nature does work in pure cultures. Yet the study of pestiferous organisms isolated from their natural environment continues to be the dominant method of teaching and research. Winogradsky (1928) mentioned that when an organism is isolated from the soil and cultured in the laboratory it becomes less and less like the organism that existed in the soil. Foster (1949) pointed out that the study of fungus activities in pure culture under the highly artificial conditions of the laboratory cannot be expected to present a true picture of the metabolic performance of an organism as it occurs in nature.

There are those who say that Koch's postulates have been a deterrent to science, especially as they have been modified in plant pathology. Implicit in Koch's postulates is the notion of one causal agent—one disease. We can point with pride, however, to the accomplishments made under controlled conditions in which only one parasite is active at a time. But one may also wonder how much further along we would be in understanding the nature of diseases and their causal agents if we had heeded scientists decades ago. Actually, in plant pathology, the nematologists have done the most to promote the concept of interactions. Many pathologists have been slow in accepting such a concept. This is surprising, as a pathologist of the old school noted long ago that Fusarium wilt of cotton in Alabama was worse when root knot was present than when it wasn't (Atkinson, 1892). The pathological aspects of interactions have been reviewed by Powell (1971). Wallace (1973) also argued effectively for the multiple facet concept of disease. Some interactions among organisms that alter disease expression are summarized in Table 10.1. Much emphasis, however, has been on disease expression, and less attention has been given to population changes of the microorganisms involved.

Nematode populations are affected directly or indirectly not only by their own activities but also by other organisms, large or small. The importance of interactions with secondary invaders is becoming increasingly recognized. Tobacco plants inoculated with *Meloidogyne incognita* and then with fungi not normally pathogenic to tobacco had more root necrosis than plants inoculated with only root-knot nematodes (Powell et al., 1971). Mayol and Bergeson (1970) and Starr and Mai (1976a) found that root-knot nematodes caused more damage in natural soil than in autoclaved soil. Although nonsterile soil might contain other pathogenic organisms that could cause damage independently, there was no evidence of this in the work by Mayol and Bergeson.

NEMATODES AND FUNGI

Most work on interactions has been with fungi and nematodes (Table 10.1). Early discoveries of such interactions were mostly of interrelationships between root-knot nematodes and Fusarium wilts. Lists now include all major groups of parasitic fungi and a diversity of nematodes. New associations doubtless will be found; interaction phenomena can be accepted as a frequent and probably even normal occurrence. It is now necessary to clarify the effects of these associations on nematode populations in the field.

If a plant is killed or severely stunted by a fungus, food sources for nematodes will be reduced or eliminated and the nematode population will diminish. Sometimes the interaction is not merely a straightforward depletion of food. Populations of sedentary endoparasites, such as species of *Meloidogyne* and *Heterodera*, are reduced if fungi destroy plant tissue to the extent that formation of giant cells or syncytia is prevented or inhibited. Nematode feeding was disturbed when *Plasmodiophora brassicae* parasitized giant cells incited by *Meloidogyne incognita acrita* (Ryder and Crittenden, 1965). Nematode populations do not always decrease and may even increase when there is coinfection with a fungus. Population increases occurred with coinfection of *Verticillium dahliae* with *Tylenchorhynchus*

Table 10.1. Interaction of nematodes with other organisms often related to disease

Nematode	Associated organism	Host	Comments	Reference
		Algae		
Pristionchus lheritieri	Algae		Some algae can pass through nematodes and still be viable	Leake and Jensen, 1970
		Bacteria		
Anguina tritici	*Corynebacterium tritici*		*Bacterium is a contaminant on juveniles and not able to cause yellow ear rot in absence of nematode; in absence of bacteria only galling is caused. Both symptoms occur when both organisms are present.*	Gupta and Swarup, 1972
M. incognita	*Corynebacterium michiganense*	Tomato	Canker worse with nematode in some varieties and inoculation conditions; bacteria had no effect on reproduction of nematode	De Moura et al., 1975
Aphelenchoides fragariae	*Xanthomonas begoniae*	Begonia	Bacterial leaf spot more severe in presence of nematode	Riedel and Larsen, 1974
A. ritzemabosi	*Corynebacterium fascians*	Strawberry	In combination, "cauliflower disease" is produced; either alone produces separate symptoms.	Pitcher and Crosse, 1958

Nematode	Pathogen	Host plant	Effect	Reference
Criconemoides xenoplax	*Pseudomonas syringae* and waterlogging	Peach	Nematode increased susceptibility of peach to both; no interaction with *Pythium*	Lownsbery et al., 1973
C. xenoplax	*Pseudomonas syringae*	Plum	Bacterial cankers more extensive on trees infected with nematodes	Mojtahedi et al., 1975
Ditylenchus dipsaci	*Corynebacterium insidiosum*	Alfalfa	Nematode can transmit the bacteria.	Hawn, 1963
D. dipsaci	*Corynebacterium insidiosum*	Alfalfa	Resistance to bacterial wilt can be "broken" by the nematode	Hawn and Hanna, 1967
Meloidogyne hapla	*Agrobacterium tumefaciens*	Red raspberry	Crown galls result only when plants are inoculated with *M. hapla*	Griffin et al., 1968
M. hapla	*Corynebacterium insidiosum*	Alfalfa	Worse with combination; varied with variety but not temperature	Griffin and Hunt, 1972
M. hapla	*Corynebacterium insidiosum*	Alfalfa	Increased incidence of wilt	Hunt et al., 1971
M. hapla	*Corynebacterium insidiosum*	Alfalfa	Increased wilt, also increased crown and root rot in absence of bacterial wilt	Norton, 1969
M. javanica	*Pseudomonas marginata*	Gladiolus	Presence of nematode increased incidence and severity of scab on the corms	El-Goorani et al., 1974
Meloidogyne spp.	*Pseudomonas caryophylii*	Carnation	Nematodes increased rate of wilting; *Xiphinema diversicaudatum* had no effect	Stewart and Schindler, 1956

Table 10.1. (Continued)

Nematode	Associated organism	Host	Comments	Reference
Helicotylenchus multicintus and *Pratylenchus pratensis*	*Erwinia caratovorum, Xanthomonas phaseoli, Pseudomonas fluorescens*		Nematodes act as vector for bacteria	Kalinenko, 1936
Pristionchus lhertieri	*Agrobacterium tumefaciens, Erwinia amylovora, E. carotovora,* and *Pseudomonas phaseolicola*		Bacteria survived passage through alimentary canal of nematode	Chantanao and Jensen, 1969
Helicotylenchus dihystera	*Pseudomonas carophylii*	Carnation	Increased incidence of bacterial wilt	Stewart and Schindler, 1956
		Fungi		
Aglenchus agricola	*Fusarium roseum*	Corn	Increase potential of fungus; not true with *Pythium ultimum*	Kisiel et al., 1969
Aphelenchoides cibolensis, A. composticola	*Armillaria mellea*	*Pinus ponderosa*	Nematodes suppressed development of *A. mellea* and reduced mortality of seedlings	Riffle, 1973
A. parietinus, Aphelenchus avenae, Acrobeles butchlii, Cephalobus elongatus	*Fusarium moniliforme*	Cotton	Combined nematodes reduced plant weight relative to fungus alone	Christie, 1937
Aphelenchus avenae	*Aspergillus niger*	In vitro	Nematode greatly immobilized in culture filtrate of *A. niger* but not *Fusarium* sp.	Mankau, 1969

Nematode	Fungus	Crop	Interaction	Reference
A. avenae	*Pythium arrhenomanes*	Corn	*Pythium* root rot decreased when nematode added	Rhoades and Linford, 1959
Belonolaimus longicaudatus	*Fusarium oxysporum* f. *vasinfectum*	Cotton	Nematode increases wilt in resistant and susceptible varieties	Holdeman and Graham, 1954
Helicotylenchus dihystera	*Phytophthora cinnamomi*	*Pinus echinata*	Nematode created infection courts for fungus	Barham et al., 1974
H. dihystera	*Pythium graminicola*	Sugarcane	Positive interaction on top growth, but not on root	Apt and Koike, 1962a
Heterodera avenae	*Rhizoctonia solani*	Wheat	In combination, greater reduction in tillering, height, weight, root number, and length than with either alone	Meagher and Chambers, 1971
H. rostochiensis	Grey sterile fungus	Tomato	Fungus probably prevents syncytium formation by the nematode	Roy, 1968
H. schachtii	*Fusarium oxysporum*	Sugar beet	Damage was less when fungus was present; fungus inhibited nematode invasion	Jorgenson, 1970
H. schachtii	*Pythium ultimum*	Sugar beet	Synergistic on damping off when in combination	Whitney, 1974
H. tabacum	*Fusarium oxysporum* f. *lycopersici, Verticillium albo-atrum*	Tomato	More *Verticillium* but less *Fusarium* wilt in the presence than absence of *H. tabacum*	Miller, 1975
Meloidogyne arenaria	*Aspergillus flavus*	Peanut	*M. arenaria* damage did not affect incidence of fungus or aflatoxin contamination to peanuts	Bell et al., 1971
M. arenaria, M. incognita, M. javanica	*Fusarium oxysporum, F. nicotianae*	Tobacco	Significant infection occurred only when nematodes and fungus were both present	Porter and Powell, 1967

Table 10.1. (Continued)

Nematode	Associated organism	Host	Comments	Reference
M. hapla	Aspergillus flavus	Peanut	Greater amount of fungus in kernel but not shell when both are present than with fungus alone	Minton et al., 1969
M. hapla	Fusarium oxysporum f. lycopersicon	Tomato	Increases susceptibility to wilt	Jenkins and Coursen, 1957
M. hapla	Verticillium dahliae	Strawberry	Control of root knot delayed onset of wilt symptoms and reduced severity of disease	Meagher and Jenkins, 1970
M. incognita	Pythium ultimum, Curvularia trifolii, Botrytis cinerea, Aspergillus ochraceus, Penicillium martensii, Trichoderma harzianum	Tobacco	Nematode predisposes plant roots so that generally nonpathogenic fungi cause increased root necrosis	Powell et al., 1971
M. incognita	Fusarium oxysporum f. nicotianae, Alternaria tenius	Tobacco	Brown spot of leaves was most severe when all three pathogens occurred together; no interaction in absence of nematode	Powell and Batten, 1969
M. incognita	Fusarium oxysporum f. melonis	Muskmelon	Nematode reduced resistance in Saticoy variety of muskmelon	Bergeson, 1975
M. incognita, M. incognita acrita, Trichodorus sp., Tylenchorhynchus sp., Helicotylenchus sp.	Fusarium oxysporum f. vasinfectum	Cotton	Only Meloidogyne spp. significantly increased incidence of wilt	Martin et al., 1956

Nematode	Fungus	Host	Interaction	Reference
M. incognita, M. incognita acrita	*Phytophthora parsitica f. nicotiania*	Tobacco	Nematodes increase severity of black shank; fungus has affinity for hypertrophied and hyperblastic areas of tissue	Powell and Nusbaum, 1960
M. incognita acrita	*Fusarium moniliforme*	Corn	Seedling dry weight less with both than with either alone	Palmer and MacDonald, 1974
M. incognita, Belonolaimus longicaudatus	*Pythium aphanidermatum*	Chrysanthemum	Disease worse and appeared earlier with either nematode plus fungus than with nematode alone; fungus suppressed egg production of *M. incognita*	Johnson and Littrell, 1970
M. incognita	*Rhizoctonia solani*	Tobacco	Worse when *M. incognita* precedes *R. solani* by 10 or 21 days than when introduced simultaneously or separately	Batten and Powell, 1971
M. incognita	*Rhizoctonia solani*	Cotton	Synergism increased in cotton as coarse particle soil increased	Carter, 1975
M. incognita acrita	*Fusarium oxysporum f. conglutinans*	Cabbage	Nematode did not affect amount of wilt in resistant varieties but made a susceptible variety more susceptible	Fassuliotis and Rau, 1969
M. incognita acrita	*Pythium graminicola*	Sugarcane	Positive interaction on top growth but not on root growth	Apt and Koike, 1962b
M. incognita acrita	*Alternaria tenuis, Fusarium oxysporum f. sp. vasinfectum, Glomerella gossypii, Rhizoctonia solani*	Cotton	Disease worse with nematodes than with fungus alone; nematodes had no effect with some other fungi	Cauquil and Shepherd, 1970

Table 10.1. (Continued)

Nematode	Associated organism	Host	Comments	Reference
M. incognita acrita, Radopholus similis	Fusarium oxysporum f. cubense	Banana	Incubation period of fungus shortened by presence of nematodes	Loos, 1959
M. incognita acrita	Rhizoctonia solani	Cotton	Cotton seedling disease was increased in presence of nematode	Reynolds and Hanson, 1957
M. incognita acrita	Verticillium albo-atrum	Cotton	Nematodes inoculated before and after emergence increased incidence and severity of wilt	Khoury and Alcorn, 1973
M. incognita	Secondary invaders	Tomato	Secondary invaders decreased top and root weight	Mayol and Bergeson, 1970
Meloidogyne sp.	Fusarium oxysporum f. lycopersici	Tomato	Nematode reduces resistance but not immunity to wilt	Harrison and Young, 1941
Meloidogyne sp.	Fusarium oxysporum f. gladioli	Gladiolus	Nematode had very little effect on incidence of Fusarium infection	McClellan and Christie, 1949
	Fusarium oxysporum f. lycopersici	Tomato		
	Fusarium solani f. cucurbitae	Squash		
Meloidogyne spp. (six varieties or species)	Fusarium oxysporum f. nicotianae	Carnation	Synergistic effect only with root knot plus fungus	Schindler et al., 1961
Meloidogyne spp.	Phytophthora parasitica f. nicotianae	Tobacco	Increase in incidence and severity of black shank in presence of nematode; more than just wounding effect	Sasser et al., 1955

Nematode	Fungus	Plant	Effect	Reference
M. javanica	Fusarium oxysporum f. lycopersici	Tomato	Fungus more numerous in rhizosphere of root-knot-infected plants than with no root knot; combined infections increased fungus in roots but not foliage	Bergeson et al., 1970
M. javanica	Fusarium oxysporum f. tracheiphilum	Cowpea	Wilt more severe in presence of nematode	Thomason et al., 1959
M. javanica	Macrophomina phaseoli	Ligustrum japonicum	Leaf necrosis, abscission, twig dieback, stunting, reduction and necrosis of roots, and stand were all greater on plants infected with both pathogens	Alfieri and Stokes, 1971
M. javanica	Macrophomina phaseoli	Kenaf	Incidence and severity of root rot increased with nematode	Tu and Cheng, 1971
Pratylenchus brachyurus	Phytophthora parasitica f. nicotiana	Tobacco	Nematode acts as wounding agent	Inagaki and Powell, 1969
P. brachyurus	Fusarium oxysporum f. vasinfectum	Cotton	With low inoculum, wilt incidence increased in wilt-susceptible but not wilt-resistant variety with both pathogens; nematode population increased in presence of fungus in susceptible variety	Michell and Powell, 1972
P. minyus	Verticillium dahliae	Peppermint	P. minyus increases incidence and severity of wilt when parasitizing separate root systems of same plant	Faulkner et al., 1970

Table 10.1. (Continued)

Nematode	Associated organism	Host	Comments	Reference
P. neglectus	*Rhizoctonia solani*	Wheat	When *P. neglectus* was reduced so was *R. solani*, but not vice versa	Benedict and Mountain, 1956
P. neglectus	*Verticillium dahliae* f. *menthae*	Peppermint	Presence of nematode increased incidence and severity of wilt at five temperatures; optimum temperature for nematode was 24 C with fungus, 30 C without	Faulkner and Bolander, 1969
P. penetrans, Meloidogyne spp.	*Fusarium solani*, f. sp. *phaseoli*	Red kidney bean	At low fungus inocula, more plants had dry rot when infected with either nematode than when not infected; nematodes had no effect on severity of dry rot	Hutton et al., 1973
P. penetrans	*Verticillium dahliae*	Eggplant, tomato	Reproduction of nematode increased in roots	Mountain and McKeen, 1962
P. penetrans	*Fusarium oxysporum* f. *pisi*	Pea	Nematode "broke" resistance of a pea variety to the fungus	Oyekan and Mitchell, 1971
P. penetrans	*Aphanomyces euteiches*	Pea	Nematode increases severity of root rot with low but not high levels of fungus	Oyekan and Mitchell, 1972
P. penetrans	*Verticillium albo-atrum*	Tomato	Wilt incidence increased with presence of nematode, but root population of nematodes was not increased with fungus	Conroy et al., 1972
Trichodorus christiei	*Pythium, Rhizoctonia, Fusarium*	Cotton	Several nematodes and fungi evidently involved in stunt	Bird et al., 1971

Nematode	Fungus	Host	Effect	Reference
Pristionchus lheritieri	*Fusarium oxysporum, Verticillium dahliae*		Ingested spores are protected from lethal doses of fungicides	Jensen and Siemer, 1971
Radopholus similis	*Fusarium oxysporum, F. solani*	Citrus	Higher incidence of fungi in roots invaded by nematode than when not invaded	Feldmesser et al., 1960
R. similis	*Fusarium oxysporum, F. solani*	Grapefruit	Root rot more severe in combination than when nematode or fungi used alone	Feder and Feldermesser, 1961
Rotylenchulus reniformis	*Fusarium vasinfectum*	Cotton	Increase wilt in some strains of cotton	Neal, 1954
Tylenchorhynchus capitatus	*Verticillium dahliae*	Tomato	Production of nematode increased on roots	Mountain and McKeen, 1962
T. claytoni	*Fusarium oxysporum f. nicotianae*	Tobacco	Nematode increased incidence of wilt in susceptible variety	Holdeman, 1956
T. dubius	*Fusarium roseum*	*Poa pratensis*	*Fusarium* blight much worse in presence of nematode	Vargas and Laughlin, 1972
T. martini	*Aphanomyces euteiches*	Pea	Nematode increased severity of Aphanomyces symptoms only with strongly pathogenic isolate of fungus	Haglund and King, 1961
Tylenchulus semipenetrans	*Fusarium solani*	Citrus	Yield reduction greater in combination than with either alone; varied with temperature	O'Bannon et al., 1967; Van Gundy and Tsao, 1963
Nematodes				
Belonolaimus longicaudatus, Criconemoides ornatus, Tylenchorhynchus martini		Bermuda-grass	When mixed, *B. longicaudatus* is better competitor than other two; competitive ability varied with variety	Johnson, 1970

193

Table 10.1. (Continued)

Nematode	Associated organism	Host	Comments	Reference
Helicotylenchus dihystera, Tylenchorhynchus claytoni, Trichodorus christiei, Pratylenchus brachyurus, P. zeae		Corn or soybean	Populations of *P. zeae* and *T. christiei* were higher when together than alone; *T. christiei* and *T. claytoni* were suppressed in combination with the other than alone but varied with variety; *P. brachyurus* adversely affected by *T. christiei* and *T. dihystera*	A. W. Johnson and Nusbaum, 1968
Heterodera trifolii and *Pratylenchus penetrans*		Red clover seedlings	Neither affected the invasion of the other	Freckman and Chapman, 1972
Meloidogyne hapla, Xiphinema americanum		Alfalfa	Twice as many *X. americanum* in soil in absence than in presence of *M. hapla*	Norton, 1969

M. hapla and M. javanica	Tomato	M. javanica predominated in mixed culture; M. hapla produced more terminal galls and lateral roots	Kinloch and Allen, 1972
Pratylenchus brachyurus and Meloidogyne spp.		More penetration by P. brachyurus of cotton when M. incognita prior to or with P. brachyurus; M. incognita suppressed P. brachyurus on tomato but no effect on alfalfa or tobacco; M. incognita on cotton was generally inhibited by presence of P. brachyurus	Gay and Bird, 1973
P. penetrans and Heterodera tabacum	Tobacco	Inverse relationship in numbers	Miller, 1970
Tylenchorhynchus martini, Pratylenchus penetrans	Red clover	P. penetrans increased as well with T. martini as it did alone; T. martini increased only 10–25% as much with P. penetrans as alone	Chapman, 1959

capitatus on tomato (Mountain and McKeen, 1962); *V. dahliae* with *Pratylenchus penetrans* in eggplant and tomato, but not in pepper (Mountain and McKeen, 1962); *V. dahliae* with *P. minyus* in peppermint (Faulkner and Skotland, 1965); and *Fusarium oxysporum* var. *vasinfectum* with *P. brachyurus* (Michell and Powell, 1972). Nematode populations did not increase with *P. penetrans* and *V. albo-atrum* in other tests, however (Bergeson, 1963; Conroy et al., 1972).

Several instances are known in which galled tissue is more susceptible to fungi than nongalled tissue. For example, Golden and Van Gundy (1975) found that sclerotia of *Rhizoctonia solani* formed only on galled tissue of okra and tomato roots infected with *M. incognita.* Using cellophane membranes they found that the fungus was specifically attracted to the galls.

Not all interactions result from organisms being in close proximity. Brown leaf spot of tobacco, caused by *Alternaria tenuis*, was most severe in the presence of *Meloidogyne incognita* and *Fusarium oxysporum* f. *nicotianae.* Although the nematode predisposed plants to infection by *A. tenuis*, the magnitude of the disease was less than when the *Fusarium* was also present (Powell and Batten, 1969). Numbers of *Xiphinema americanum* increased on alfalfa when foliar diseases were controlled by a fungicide. The increase was due presumably to a better root system resulting from control of foliar diseases (Norton, 1965b).

Sex ratios of nematodes also may be changed as a result of cohabitation by nematodes and fungi. These were discussed in Chapter 5.

NEMATODES AND BACTERIA

It has been suspected, and even accepted as fact in the absence of much experimental proof, that saprozoic nematodes feed on and gain much of their sustenance from bacteria. Little is known about whether these nematodes are carriers of plant pathogenic bacteria, but there is evidence that they are. Steiner (1933) suggested that *Pelodera lambdiensis* is a probable carrier of *Pseudomonas tolaasii*, a pathogen of cultivated mushrooms. *Pseudomonas syringae* was reported to survive its passage through the alimentary canal of *Pristionchus lheritieri* (Jensen, 1967). Later, Chantanao and Jensen (1969) demonstrated that *Agrobacterium tumefaciens, Erwinia amylovora, E. caratovora, Pseudomonas phaseolicola,* and the nonplant pathogen *Serratia marcescens* were viable after passage through *P. lheritieri.* Females were more efficient than males at ingesting and retaining the bacteria.

Kalinenko (1936) was the first to demonstrate that bacteria were carried through the digestive tract of a plant-parasitic nematode. He surface sterilized *Helicotylenchus multicinctus* and *Pratylenchus pratensis* and placed them in broth tubes. Of 127 nematodes so treated, 27 resulted in growth of microbes. Isolates obtained were *Erwinia carotovora, Xanthomonas phaseoli, Pseudomonas fluoroscence,* and *Bacterium necrosis* and were identical physiologically and serologically to those obtained earlier from infected tissue. He demonstrated further that these bacteria cultures were pathogenic on rubber plants. Stewart and Schindler (1956) found that some specificity is involved in a bacteria and nematode association in carnation wilt. Disease symptoms occurred when *Pseudomonas caryophylli* was associated with four *Meloidogyne* species, but no symptoms were observed when *P. caryophylli* was associated with the ectoparasites *Helicotylenchus dihystera, Xiphinema diversicaudatum,* and *Ditylenchus* sp.

Nematodes may act as vectors of pathogens either from within a plant or from the soil. Typical examples are cauliflower disease of strawberry and ear-cockle disease of wheat (Carne, 1926; Cheo, 1946; Pitcher, 1963; Sabet, 1954; Vasudeva and Hingorani, 1952). Carne (1926) was the first to report that a bacterium and nematode complex was involved in ear-cockle disease of wheat. Cheo (1946) demonstrated that the combination is necessary for the production of symptoms in some instances. Sabet (1954) demonstrated not only that the combination of bacteria and nematode is involved in the disease complex but also that these two pathogens have to be carried to the ear of wheat from the growing tip. When he placed inoculum on the aerial parts of the plant, infection did not result. In order to establish that nematodes act as vectors for the transportation of bacteria in the cauliflower disease of strawberry, Pitcher (1963) employed both pot-grown plants and aseptic seedlings grown in agar tubes amended with nutrients. Symptom expression was different when either *Corynebacterium fascians* or *Aphelenchoides ritzemabosi* was inoculated alone than when the organisms were combined.

Hawn (1963) obtained from 4 to 27% increase in wilting of alfalfa in the presence of both *Ditylenchus dipsaci* and *Corynebacterium insidiosum* over that with *Corynebacterium insidiosum* alone. He concluded that the nematodes carried the bacterium on their body surface into the crown buds. Nematodes alone did not affect the incidence of wilt on plants that were wounded first.

Nematodes and Nodule-Forming Bacteria

There are many instance in which the presence of nematodes results in reduced nodulation in plants. These associations include *Heterodera glycines* and *Rhizobium japonicum* (Barker et al., 1972) and *Meloidogyne javanica, Heterodera trifolii,* and *Rhizobium trifolii* (Taha and Raski,

1969). The latter authors found that nodule formation in white clover was not affected unless the nematodes were introduced one week after inoculations with *Rhizobium*. Invasion of nodules by *M. javanica* did not affect the efficiency of nitrogen production directly, but it did so indirectly by the earlier deterioration of the nematode-infected nodules. Later, Jatala et al. (1974) reported that *Pristionchus lheritieri* can be a carrier of *Rhizobium japonicum*, the ecological importance of which is not known.

COHABITATION BY NEMATODES

Cohabitation by nematodes has received less attention than interactions between nematodes and fungi, but they could be equally important. Nematodes often have different preferred feeding sites (Table 3.6). Theoretically, nematodes of two species feeding at different sites would not inhibit each other until indirect effects, such as gas exchange, depleted host nutrition and until possible morphological changes occurred in the host. Also, concomitant infections and nematode development are influenced by host susceptibility (Gay and Bird, 1973).

Competition is occurring if the growth rate of an organism is greater by itself than in combination with another organism (Figure 10.1). Our knowledge of competition, which has been studied little in nematology, is not on as sound an ecological basis as it is with other organisms. Probably the same principles apply even though different mechanisms may exist. In spite of our lack of knowledge, the effects of competition should not be

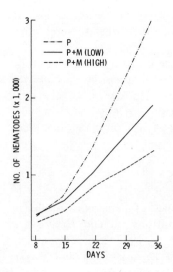

Figure 10.1   Number of *Pratylenchus penetrans* in tomato roots after 8, 15, 22, and 36 days when inoculated alone or with two densities of *Meloidogyne incognita*. P = 1000 *P. penetrans* alone; P + M (Low) = 1000 *P. penetrans* plus *M. incognita;* P + M (High) = 1000 *P. penetrans* plus 3325 *M. incognita*. (After Estores and Chen, 1972.)

underestimated. Gause's principle (Gause, 1934), or the principle of competitive exclusion, essentially states that only one species eventually can occupy a niche. In other words, if two species occupy a niche, one eventually will outcompete the other even though each species by itself could function in that niche. The competitive interaction, however, need not involve predation, parasitism, or disease but is a result of differences in reproductive capacities, amount of food consumed, and so on, among competing species. As much attention as Gause's principle has received on theoretical and experimental grounds, it sometimes is difficult to apply with nematodes under natural conditions. First, what is a niche, and how much of a change is necessary before a second niche can be called different? Second, the heterogeneity of the soil provides abundant niches for many species of nematodes. Third, due to the growth and death of root portions or of entire root systems, many niches are ephemeral, existing for an insufficient time for a nematode population to stabilize. Still, we often find that one species predominates in a sample, which may suggest that it is outcompeting others; or were other species excluded because of climatic, edaphic, or other extraneous factors? One of the most uniform habitats where Gause's principle might be applied in the natural world of plant nematodes is in the mesophyll tissue of leaves inhabited by more than one species of *Aphelenchoides*. The life cycle of the bud and leaf nematodes is relatively rapid, however, as is the progress of the disease and it is possible that even there is insufficient time to carry a study to completion.

There is heterogeneity in niches, and therefore species can cohabit. There are many examples in which species can be in close proximity to one another, basically inhabiting different niches, but still influence each other, directly or indirectly. Let us examine some examples.

Ectoparasites

Although there is some evidence that two different ectoparasitic species can be competitive, we do not know how extensive this is. Theoretically, if one ectoparasite feeds on root hairs while another feeds on cortical parenchyma outside of the piliferous zone, there would not be any purely physical spatial competition. However, there could be competition for nutrients, moisture, or gases that might be translocated from one feeding site to another. Buildup of toxic substances conceivably could affect nematodes many cells distant from the feeding site. Accumulation of carbon dioxide resulting from respiration of several nematodes may inhibit others. Thus there are many ways in which competition might manifest itself. I suspect that physical competition for space is seldom a limiting factor for population density but that density limitation is due to other causes.

Johnson (1970), using bermudagrass in greenhouse tests, found that *Criconemoides ornatus* was reduced significantly by either *Belonolaimus longicaudatus* or *Tylenchorhynchus martini*, the latter being the more suppressive. Populations of *Belonolaimus longicaudatus* were suppressed in combination with other nematodes in only one of six host varieties tested (Figure 10.2). *Criconemoides ornatus* or *B. longicaudatus*, alone or in combination, suppressed *T. martini* on two host varieties. *B. longicaudatus* was generally a better competitor than either *C. ornatus* or *T. martini*. The fact that, in Johnson's work, the populations of three species of nematodes alone or in combination did not behave similarly on six varieties of hosts is important although not surprising. Populations of a nematode were sometimes larger, although not significantly so, after 155 days in the presence of a second species of nematode than when alone. Likewise a marked but not significant yield increase occurred in one variety when all three nematodes were present compared with the noninfested treatment. It was not reported, however, whether the terminal populations at 155 days were the peak populations.

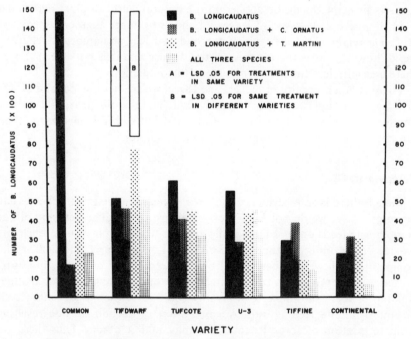

Figure 10.2. Population of *Belonolaimus longicaudatus* for each of six varieties of bermudagrass after 155 days with *B. longicaudatus, Criconemoides ornatus,* and *Tylenchorhynchus martini* singly and combined. (From A. W. Johnson, 1970.)

Ecto- and Endoparasites

When an ectoparasite and a migratory endoparasite were inoculated on alfalfa or red clover in 1:1 ratios, *Pratylenchus penetrans* increased as well in combination with *Tylenchorhynchus martini* as it did alone. However, *T. martini* in combination with *P. penetrans* increased only 10 to 25% as much as it did alone (Chapman, 1959). *Tylenchorhynchus agri* inhibited increase of *Meloidogyne naasi* in creeping bentgrass (Sikora et al., 1972). Creeping bentgrass was a poor host for *P. penetrans*, and the nematode did not inhibit *M. naasi*. Twice as many *Xiphinema americanum* were associated with alfalfa after 4 years in plots not infested with *Meloidogyne hapla* as in infested plots (Norton, 1969). Morphological changes in the root due to infection by *M. hapla* are suspected of reducing feeding sites for *X. americanum*, but depleted host nutrition may also be a factor.

Migratory and Sedentary Endoparasites

Simultaneous inoculation of red clover with *Heterodera trifolii* and *P. penetrans* did not affect the amount of invasion of either. Neither did 120-hour prior invasion by *H. trifolii* influence subsequent invasion by *P. penetrans* (Freckman and Chapman, 1972). *Heterodera tabacum*, however, suppressed infection and survival of *P. penetrans* and *Tylenchorhynchus claytoni*, but the former was affected more. Infection by *H. tabacum* was reduced when soils containing cysts were mixed with soils containing *P. penetrans* or *T. claytoni* (Miller and Wihrheim, 1968). *Heterodera tabacum* increased slowly in the presence of large numbers of *P. penetrans* and increased most rapidly in the presence of few *P. penetrans*, a nematode that is uncommon in tobacco fields heavily infested with *H. tabacum* (Miller, 1970).

Several studies have been made relative to cohabitation of *Pratylenchus* spp. and *Meloidogyne* spp. These nematodes have different invasion and feeding sites, and any interaction between the two is apt to be indirect. Estores and Chen (1972) found that densities of *Pratylenchus* and *Meloidogyne incognita* were smaller in combination on tomato roots than when they were in monoculture. There was little difference in populations when inoculations of the two species were 10 days apart. Root penetration by *P. penetrans*, however, was inhibited by the presence of *M. incognita*. While the inhibition of *P. penetrans* by *M. incognita* involved factors other than available feeding sites, the inhibitors of *M. incognita* by *P. penetrans* was governed largely by the amount of root tissue on which *M. incognita* could feed.

Turner and Chapman (1972) obtained results that seem contradictory to those of Estores and Chen, but this may be a reflection of differences in hosts, experimental conditions, and length of experiments. Turner and Chapman found that invasion of alfalfa or red clover by *Meloidogyne incognita* or *Pratylenchus penetrans* was not affected when 50 or fewer individuals or each species of nematode were inoculated simultaneously. When 50 *M. incognita* and 200 *P. penetrans* were inoculated simultaneously on alfalfa, invasion by *M. incognita* was greatly reduced. Reciprocal combinations had no effect on invasion by either nematode. Invasion of red clover was not affected by any combination tested. Invasion by 50 *M. incognita* was also reduced when 200 *P. penetrans* were used previously as inoculum, but reciprocal combinations had no effect. Populations of *Pratylenchus brachyurus* were not affected adversely by *Meloidogyne incognita* in some tobacco lines but they were in others (Johnson and Nusbaum, 1970).

Sedentary Endoparasites

It is in this category that there is often direct competition for invasion and feeding sites in a limited area, and it might be expected that one species might suppress another. Cohabitation of two species of root-knot nematodes in tomato resulted in *M. javanica* predominating other *M. hapla*. Invasion by *M. hapla* was more density-dependent than invasion by *M. javanica* (Kinloch and Allen, 1972). *Meloidogyne incognita* also suppressed *M. hapla* in two tobacco lines, and it was speculated that this might be related to the rapid necrosis of the root tips throughout the feeding areas of *M. incognita* juveniles, thus reducing infection sites for *M. hapla* (Johnson and Nusbaum, 1970).

Thus it can be seen that competition between species of nematodes exists and it is possible that some nematodes species are not able to survive and persist in a particular habitat due to, in part, their inability to compete with other species. It should be reiterated, however, that, especially in the field, an example that appears to be competition may really be differential habitation resulting from environmental alterations. It is difficult to make conclusions relative to consistency of competitiveness. In some instances, the endoparasites predominate over the ectoparasites and the sedentary endoparasites over the migratory endoparasites. There are exceptions, but our knowledge is limited.

NEMATODES AND VIRUSES

Ever since the discovery by Hewitt et al. (1958) that *Xiphinema index* can transmit a plant virus, much investigation concerning this phenomenon has been performed with other nematodes. So far, only species of *Longidorus*,

Xiphinema, and *Trichodorus* have been shown conclusively to transmit plant viruses. There are good reviews on the subject (Taylor, 1971). Important aspects relative to this treatment are (1) what is the persistence of viruses in nematodes from the standpoint that nematodes may act as survival agents for viruses; (2) does the ecology of the nematode affect the distribution of the virus either locally or regionally; and (3) is there any effect of the virus on the nematode that might affect nematode populations, directly or indirectly?

Periods of virus survival in nematodes vary with the virus, nematode, and associated plant, but they have been surprisingly long. The grapevine fanleaf virus is known to persist for up to 8 months in *Xiphinema index* in the absence of a nematode host (Taylor and Raski, 1964). The tobacco ringspot virus was transmitted by *X. americanum* after this nematode had a starvation period of 49 weeks (Bergeson et al., 1964), and McGuire (1973) found TRSV could be transmitted after storage in *X. americanum* for 9 months at O C. Apparently viruses do not survive in *Longidorus* as long as they do in *Xiphinema.*

If the viruses that are transmitted by nematodes are also transmitted by insects, weeds, or other agents, then the role of nematodes in the general distribution of the virus is probably minor. This does not negate the importance of nematodes, however, as an agent for survival in a local situation. If a nematode-transmitted virus is distributed locally, however, then the distribution of the nematode may govern the distribution of the virus if other transmitting agents are not operative. Edaphic factors that enter into the ecology of the nematode may be important. These soil factors should be examined where such a situation appears to exist.

As far as is known, however, the viruses that are transmitted by nematodes are not infectious to the nematode carrier. Some viruses may be just a passive contamination of the nematode's stylet. On the contrary, there is evidence that *X. index* survives longer when it contains the grape fanleaf virus than when it doesn't (Das and Raski, 1969). There was no evidence that the virus has any effect on nematode reproduction or that it is transmitted through nematode eggs. Population density of *Aphelenchoides ritzemabosi* and *Ditylenchus dipsaci* was less in plants infected with tobacco mosaic virus, however, than in virus-free plants. It was suggested that the reduction in nematode reproduction was the result of altered physiology of the plant due to virus infection (Weischer, 1969). In other work, *Meloidogyne javanica* grew more rapidly in tomatoes infected with tobacco mosaic virus than in noninfected ones. Whether or not the plant was infected with TMV had no effect on the number of nematodes invading the roots, but there was more nematode invasion of plants infected with tobacco ring spot virus than of the controls (Bird, 1969).

CHAPTER 11

EFFECTS OF AGRICULTURAL PRACTICES ON NEMATODE POPULATIONS

Pest management, a term that is currently in vogue, essentially means pest control. In recent years methodology has advanced greatly through the use of computers, but the same principles apply today as in former times; the aim is to minimize disease by reducing the amount or the influence of pest propagules. Resistance, use of chemicals, soil fertility practices, cultivation, and rotation, among others, are all capable of minimizing the effects of nematodes. Although one or more practices have been recommended for a given nematode, their use is not always economically feasible for a given area (Barry et al., 1974; Meagher and Rooney, 1966). Resistant varieties, if available, might not be adapted to an area. Other crops used in rotation, while reducing nematode populations, might not provide sufficient income to make planting them practical for the farmer. These aspects are reviewed by Khan (1972) and others. Unfortunately, perhaps, there has been too much reliance on chemical control. As Barry et al. (1974) point out, cultural practices should be examined as alternatives to other methods. Many cultural practices have been recommended in nematology and other disciplines. Some are difficult to implement, but there are few instances in which the farmer can do nothing. Unfortunately our knowledge about the influence of cultural practices on nematode populations has not received the attention that chemical control has. Lack of scientifically sound recommendations might be due more to ignorance than to lack of feasibility. Frequently, no one method of control will be entirely satisfactory, and it will be necessary to implement a multifaceted approach.

CROPPING SYSTEMS

A host is all important in determining whether or not a population of parasitic nematodes is able to increase, and many host range studies have proved valuable in planning programs. The degrees of infection are not always the same, however, in the greenhouse as in the field, there have been conflicting reports about susceptibilities of plant species. Plants highly resistant in the greenhouse are usually highly resistant in the field, but plants labeled moderately resistant in the greenhouse may do well in the field due to extraneous factors. The proof is in field testing, and it is here that the final decisions usually are made. Valuable screening programs have been implemented for *Heterodera rostochiensis*, *H. glycines*, *Ditylenchus dipsaci*, and other species.

There is increased interest in cropping systems as a means of nematode control, and if there is not sufficient resistance available, rotation is an effective and economical control in some instances. *Heterodera rostochiensis* and *H. glycines* have narrow host ranges in the field, and these nematodes can be maintained at low levels if suitable rotation practices are implemented. Some serious nematode pests are kept under control by natural cropping sequences. For example, *Meloidogyne hapla* can cause serious decline of alfalfa in Iowa (Norton, 1969), and soybeans are a known host. But corn is practically immune to this species, and the 13 million acres of the crop in the state probably keeps this nematode under control.

Although there have been many successes with crop rotations, host susceptibilities are not clear-cut for most species. Therefore, rotation is sometimes not effective. This seems to be especially true with nematodes that do not have an intricate host–parasite relationship, as with the root-knot and cyst nematodes (Rohde, 1972). Within a given soil, cereals generally support greater numbers of migratory nematodes than row crops such as sugar beets and potatoes (Oostenbrink et al., 1956; Winslow, 1964). The dagger nematode, *Xiphinema americanum,* is one of the more damaging nematodes to alfalfa in Iowa (Norton, 1965b), and it is common in soybean and corn fields (Ferris and Bernard, 1971; Norton et al., 1971). That there might be some host differentials is suggested by rotation studies in Illinois and Georgia, where numbers of *X. americanum* were reduced in corn although the crop supported the nematode (Ferris and Bernard, 1971; Brodie et al., 1969). Large numbers of *Helicotylenchus pseudorobustus* and *Paratylenchus projectus* were supported by all or most rotations involving corn, oats, soybeans, wheat, and a legume-grass forage mixture, although significant differences occurred among cropping systems (Ferris and Bernard, 1971). In similar work, it was found that *Pratylenchus penetrans* and *P. scribneri* increased in corn and soybean plots, but *P. neglectus* was the only species to increase on wheat. *Pratylenchus hexincisus* increased in corn but increased only moderately in soybean plots (Ferris and Bernard, 1967). Unfortunately we are ignorant of the host ranges of many nematodes relative to field crops, but many species seem polyphagous.

The most comprehensive work on nematodes and crop rotations in the United States probably has been done in Georgia and is further evidence that hosts are important in determining the structure of populations of migratory nematodes (Brodie et al., 1969; Brodie et al., 1970a,b,c; A. W. Johnson et al., 1975). Corn favored increase of *Pratylenchus zeae* and suppressed *Trichodorus christiei, Helicotylenchus dihystera,* and *Xiphinema americanum* in a 6-year rotation in Georgia (Brodie et al., 1969). Alfalfa and fescue (*Festuca arundinaceae*) favored increase of *T. christiei* but suppressed *P. zeae, H. dihystera,* and *X. americanum.* Also in Georgia, A. W. Johnson et al. (1975) found that, with various rotations involving corn, cotton, peanuts, and soybeans, monocropping peanuts was the most effective in reducing population size of most parasitic nematodes in the test area. No single cropping system suppressed all parasitic nematodes, however. Johnson and his coworkers demonstrated that if corn or cotton are to be grown peanuts and soybeans should be included in the rotation.

Starting from a small population of nematodes on newly cleared pine land, substantial increases of *Pratylenchus brachyurus, Xiphinema americanum, Tylenchorhynchus claytoni, Trichodorus christiei,* and *Helicotylenchus dihystera* did not occur until the third year or after. The degree of increase varied with the crop. But susceptible cover crops apparently can increase nematode populations to damaging levels for succeeding cultivated crops (Brodie et al., 1970b). Good et al. (1973) also found that crops varied in their ability to suppress nematodes, and this along with the works of Brodie et al. (1970a,b,c), A. W. Johsnon et al. (1975), and others indicate that rotation systems can be worked out to control nematodes in a given area. Programs must be implemented for each agronomic area in the country.

CULTIVATION

Since soil was first broken with the plow, cultivation has had a drastic impact on nematode populations. Cultivation takes many forms in crop management. It may range from practically no disturbance of the soil to frequent churnings, depending on the crop, soil requirements, and other husbandry practices such as cultivating for weed control. Some practices undoubtedly benefit nematodes while others are detrimental. Unfortunately, relatively few publications exist concerning the effects of cultivation on nematode populations and its consequences on disease. Frequent cultivations of land may be one of the best nematode control measures known, however. During cultivation, nematodes are dispersed and many die because they are unable to find food. In addition, protective debris is broken up, with the result that many nematodes cannot survive the soil environment.

Fallowing

It has been recognized for a long time that fallowing is an effective method of reducing nematode numbers in many instances. Contingent upon effectiveness, however, is the length of fallow and the temperature and moisture content of the soil. Because weeds are a good host for many nematodes, perhaps the target ones, these also should be eliminated by an herbicide or by cultivation. Cultivation not only kills weeds but, depending upon the type of cultivation, may provide greater aeration and thus promote hatching if sufficient moisture is present. In the absence of food, the juveniles die and the population decreases. Watson (1921) used this theory in attempts to control root knot in Florida by fallowing at different seasons and for various periods up to 6 months. Population size was reduced, but in no case was eradication approached.

Inverting the soil exposes the deeper layers to the heating and drying effects of the sun (Godfrey, 1947). Soil inversion is more effective in warmer than in cooler climates. In some instances, however, fallowing might not be as effective as hoped unless it is continued for long periods. Stresses for short times may induce nematodes to go into a dormant state, and survival may actually be enhanced. While testing the effects of fallowing on a mixture of *Meloidogyne javanica* and *M. hapla* at Victoria, Australia, Sauer and Giles (1957) found that three years of clean fallow reduced nematode infestation of tomato significantly, but yield increases could not be attributed to reduced nematode infestation. When the fallowing cycle was broken by planting susceptible tomatoes, nematode populations increased rapidly. The authors concluded that susceptible crops would succeed only during the first year. They also found that a summer fallow followed by a resistant winter barley crop gave the same results when the land was planted to a susceptible tomato as with completely clean fallow. *Meloidogyne incognita* and *Pratylenchus zeae* were found to survive fallowing for one year in Tennessee, although the plots were not completely weed-free late in the season (Southards, 1971). One year of summer fallow reduced the amount of *M. incognita acrita* and increased yields as much as a good nematicide (Reynolds and O'Bannon, 1966). Fallowing was effective in reducing numbers of *Heterodera avenae* in Victoria, Australia (Meagher and Rooney, 1966). Fallowing in Trinidad for 16 weeks reduced numbers of *Rotylenchulus reniformis* by 79%, but planting previously with corn was as successful as fallowing in reducing nematode populations (Brathwaite,

1974). Winter barley was as effective as clean fallow in reducing *M. incognita* in cotton in Arizona (Carter and Nieto, 1975). Short term fallowing may have beneficial effects, however, if applied during periods of nematode stress as demonstrated by Brodie and Murphy (1975). They found that a 6-week fallow followed by a cover crop was as effective in reducing populations of *Pratylenchus brachyurus, Trichodorus christiei,* and *Meloidogyne incognita* as a 9-month fallow.

Thus it seems that fallowing might have a place under special circumstances but that it has its limitations where land use for crop production is high and where water is adequate. Leaving land fallow for extended periods allows for increased soil erosion, as well as possible detrimental altering of the soil structure. In view of the results that highly resistant or immune crops seemingly serve equally well as fallowing in reducing nematode populations, the use of such crops is preferable where culturally and edaphically feasible.

Subsoiling

Subsoiling, or breaking up of the soil below the plow layer, has been investigated more extensively with root diseases caused by fungi than with nematodes. Yield increases often result after subsoiling; some are attributed to increased root volume and depth and not to reduced infection. Cotton yields increased in plots infested with either *Hoplolaimus columbus* or *Meloidogyne incognita* and subsoiled at 14 inches under the row (Bird et al., 1974). Numbers of *H. columbus* but not *M. incognita* were reduced by subsoiling. The authors postulated that the effects of nematodes might be exhibited earlier as a result of moisture stress due to compaction where there was no subsoiling than where there was no moisture stress on the plant because of subsoiling. During one year, subsoiling under the row resulted in higher yields than did fumigation. Since populations of *M. incognita* were not markedly reduced by subsoiling, nematicide treatments might be necessary in subsequent years. Hussey (1977) found that annual subsoiling was essential for maximum cotton yields and that combining a nematicide with subsoiling for control of *H. columbus* did not increase yields. The number of nematodes was reduced in the top 20 cm with subsoiling alone. Hussey attributes this to a reduction of roots in the upper zone of subsoiled plots compared with nonsubsoiled plots where roots are confined to more shallow depths by the plow pan.

Similar results were obtained in soybean plots containing *M. incognita* in that yield increases were obtained by subsoiling (Minton and Parker, 1975). Although fumigation without subsoiling increased yields, the results were not addictive when the two treatments were used together. Subsoiling reduced galling in the nonfumigated plots but not in the fumigated ones. Subsoiling had no effect on juvenile populations at the three levels (0 to 20, 20 to 33, 33 to 46 cm) studied.

Conservation Tillage

Because of soil and energy conservation, among other concerns, greater attention is being given "conservation tillage," also called "reduced" or "minimum" tillage, in many areas of the United States. Such methods, compared with conventional plowing, have increased crop yields in some instances and reduced them in others. We know little about how these practices affect nematode populations, however. Any program that provides nematodes greater protection in unbroken root debris or reduces the dilution of densities through plowing must be looked upon with concern. Until we know whether or not the practice is detrimental or beneficial to the host or parasite, and whether any effects of nematode increase are offset by increased benefits through weed control or plant vigor, it is best to take a conservative attitude in making recommendations for reduced tillage.

Fall-tilled plots, compared with plots that were not fall tilled, had reduced root knot (*Meloidogyne incognita*) when planted with corn or tobacco the following season (Southards, 1971). The susceptibility of the host plant, however, had a greater effect than tillage on nematode populations. Southards thought that fall tillage alone could not give adequate nematode control in his situation in Tennessee but that it could be employed with crop rotation or chemical treatments. On the other hand, populations of *Pratylenchus zeae* and *Helicotylenchus* spp. did not differ significantly in conventional till and no-till treatment in corn (All et al., 1977).

In 3-year tests, migratory plant-parasitic nematodes were usually fewer in soils treated with an herbicide and drilled with wheat without plowing than in plowed plots (Corbett and Webb, 1970). Differences varied with sampling time, however, and often were not great. Species of *Pratylenchus* were more affected than those of *Tylenchorhynchus* and *Helicotylenchus. Pratylenchus crenatus* were more numerous in direct seeded plots, but the numbers were equal to those of *P. minyus* in plowed plots.

When nematode populations were monitored in seven different tillage regimes in central Iowa, numbers of *Pratylenchus* in the roots were higher in the early part of the season when there was no tillage on ridged rows, compared with other treatments (Thomas, 1978) (Figure 4.1). Soil population size of *Helicotylenchus pseudorobustus* was generally higher in the no till-ridge treatment. Population fluctuations were more erratic in the no till-ridge treatment, probably due to the greater fluctuations in temperature and moisture in the ridge beds than in the flatter ones.

GRAZING

Stocking of land could alter nematode populations indirectly through changes in the plant physiology and sociology. Actually, few data are available on the effect of grazing on nematode populations. Numbers of *Pratylenchus* in the soil and roots, and numbers of *Heterodera trifolii* juveniles, members of the Mononchidae, and total nematodes in the soil were significantly increased when an average of 22.2 sheep per hectare were stocked compared with 14.8 sheep per hectare (Yeates, 1976). There were no significant differences between treatments with members of *Paratylenchus, Helicotylenchus, H. trifolii* cysts, or the *Dorylaimida*.

Smolik and Rogers (1976) found no significant differences in total nematode population densities or biomass between grazed and nongrazed plots in the Arid Lands Ecology Reserve in south-central Washington. This contrasted with findings of a study in the mixed-grass plains of South Dakota, where the total nematode biomass was higher in the nongrazed treatment (Smolik, 1974).

REFERENCES

Adamo, J. A., C. P. Madamba, and T. A. Chen. 1976. Vertical migration of the rice white-tip nematode, *Aphelenchoides besseyi*. J. Nematol. 8:146–152.

Adams, R. E., and J. J. Eichenmuller. 1963. A bacterial infection of *Xiphinema americanum*. Phytopathology 53:745. (Abstr.)

Alfieri, S. A., Jr., and D. C. Stokes. 1971. Interaction of *Macrophomina phaseolina* and *Meloidogyne javanica* on *Ligustrum japonicum*. Phytopathology 61: 1297–1298.

All, J. N., C. W. Kuhn, R. N. Gallaher, M. D. Jellum, and R. S. Hussey. 1977. Influence of no-tillage-cropping, carbofuran, and hybrid resistance on dynamics of maize chlorotic dwarf and maize dwarf mosaic diseases of corn. J. Econ. Entomol. 70:221–225.

Allen, M. W. 1955. A review of the nematode genus *Tylenchorhynchus*. Univ. Calif. Pub. Zool. 61:129–166.

Allen, M. W. 1957. A review of the nematode genus *Trichodorus* with descriptions of ten new species. Nematologica 2:32–62.

Andrássy, I. 1956. The determination of volume and weight of nematodes. Acta. Zool. (Hungarian Acad. Sci. 2(1–3):1–15) *in* B. M. Zuckerman, M. W. Brzeski, and K. H. Deubert (Eds). English translation of selected East European papers in nematology. University of Massachusetts. 1967.

Andrássy, I. 1963. Freilebende Nematoden aus Angola, I. Einige moosbewohnende Nematoden. Publiçacões Culturais Companhia Diamantes de Angola. Lisbon 66:55–79.

Andrewartha, H. G. 1970. Introduction to the study of animal populations. Methuen, London.

Andrewartha, H. G., and L. C. Birch. 1954. The distribution and abundance of animals. University of Chicago Press, Chicago.

Apt, W. J., and H. Koike. 1962a. Pathogenicity of *Helicotylenchus nannus* and its relation with *Pythium graminicola* on sugarcane in Hawaii. Phytopathology 52:798–802.

Apt, W. J., and H. Koike. 1962b. Pathogenicity of *Meloidogyne incognita acrita* and its relation with *Pythium graminicola* on sugarcane in Hawaii. Phytopathology 52:1180–1184.

Arndt, C. H., and J. R. Christie. 1937. The comparative role of certain nematodes and fungi in the etiology of damping off, or soreshin, of cotton. Phytopathology 27:569–572.

Atkinson, G. F. 1892. Some diseases of cotton. Ala. Agric. Expt. Sta. Bull. 41:1–65.

Baines, R. C. 1974. The effect of soil type on movement and infection rate of larvae of *Tylenchulus semipenetrans*. J. Nematol. 6:60–62.

Baker, A. D. 1957. Notes on some nematodes in Canada, 1956. Can. Insect Pest Rev. 35:120–122.

Baker, K. F., and R. J. Cook. 1974. Biological control of plant pathogens. Freeman, San Francisco.

Banage, W. B. 1963. The ecological importance of free-living soil nematodes with special reference to those of moorland soil. J. Animal Ecol. 32:133–140.

Banage, W. B., and S. A. Visser. 1965. The effect of some fatty acids and pH on a soil nematode. Nematologica 11:255–262.

Barham, R. O., D. H. Marx, and J. L. Ruehle. 1974. Infection of ectomycorrhizal and nonmycorrhizal roots of shortleaf pine by nematodes and *Phytophthora cinnamomi*. Phytopathology 64:1260–1264.

Barker, K. R., and C. J. Nusbaum. 1971. Diagnostic and advisory programs. Pp. 281–301 *in* B. M. Zuckerman, W. F. Mai, and R. A. Rohde, Eds., Plant Parasitic Nematodes. Vol. I. Academic, New York.

Barker, K. R., and J. N. Sasser. 1959. Biology and control of the stem nematode, *Ditylenchus dipsaci*. Phytopathology 49:664–670.

Barker, K. R., C. J. Nusbaum, and L. A. Nelson. 1969a. Seasonal population dynamics of selected plant-parasitic nematodes as measured by three extraction procedures. J. Nematol. 1:232–239.

Barker, K. R., C. J. Nusbaum, and L. A. Nelson. 1969b. Effects of storage temperature and extraction procedure on recovery on recovery of plant parasitic nematodes from field soils. J. Nematol. 1:240–247.

Barker, K. R., D. Huisingh, and S. A. Johnston. 1972. Antagonistic interaction between *Heterodera glycines* and *Rhizobium japonicum* on soybean. Phytopathology 62:1201–1205.

Barrows, K. C. 1939. Studies on the nature of root-knot resistance. J. Agr. Res. 58:263–271.

Barry, E. R., R. H. Brown, and B. R. Elliott. 1974. Cereal cyst nematode (*Heterodera avenae*) in Victoria: Influence of cultural practices on grain yields and nematode populations. Austral. J. Exper. Agr. Animal Husb. 14:566–571.

Bastian, H. C. 1865. Monograph on the Anguillulidae, or free nematoids, marine, land, and freshwater; with descriptions of 100 new species. Trans. Linn. Soc. Lond. 25:73–184.

Batten, C. K., and N. T. Powell. 1971. The *Rhizoctonia-Meloidogyne* disease complex in flue-cured tobacco. J. Nematol. 3:164–169.

Baxter, R. I., and C. D. Blake. 1968. *Pratylenchus thornei*—a cause of root necrosis in wheat. Nematologica 14:351–361.

Beals, E. 1960. Forest bird communities in the Apostle Islands of Wisconsin. Wilson Bull. 72:156–181.

Bell, D. K., N. A. Minton, and B. Doupnik, Jr. 1971. Effects of *Meloidogyne arenaria, Aspergillus flavus,* and curing time on infection of peanut pods by *Aspergillus flavus.* Phytopathology 61:1038–1039.

Benedict, W. G., and W. B. Mountain. 1956. Studies on the etiology of a root rot of winter wheat in southwestern Ontario. Can. J. Bot. 34:159–174.

Bergeson, G. B. 1959. The influence of temperature on the survival of some species of the genus *Meloidogyne* in the absence of a host. Nematologica 5:344–354.

Bergeson, G. B. 1963. Influence of *Pratylenchus penetrans* alone and in combination with *Verticillium albo-atrum* on growth of peppermint. Phytopathology 53:1164–1166.

Bergeson, G. B. 1975. The effect of *Meloidogyne incognita* on the resistance of four muskmelon varieties to Fusarium wilt. Plant Dis. Rep. 59:410–413.

Bergeson, G. B., K. L. Athow, F. A. Laviolette, and Sister M. Thomasine. 1964. Transmission, movement, and vector relationships of tobacco ringspot virus in soybean. Phytopathology 54:723–728.

Bergeson, G. B., S. D. Van Gundy, and I. J. Thomason. 1970. Effect of *Meloidogyne javanica* on rhizosphere microflora and Fusarium wilt of tomato. Phytopathology 60:1245–1249.

Bergman, B. H. H., and A. J. van Duuren. 1959. Sugar beet eelworm and its control. VII. The action of metabolic products of some microorganisms on the larvae of *Hecterodera schachtii.* Mededeel. Inst. Ration. Suikerprod. Bergen-O-Z. 29:27–52.

Bhatt, B. D., and R. A. Rohde. 1970. The influence of environmental factors on the respiration of plant-parasitic nematodes. J. Nematol. 2:277–285.

Bhatti, D. S., H. Hirschmann, and J. N. Sasser. 1972. Post-infection development of *Heterodera lespedezae.* J. Nematol. 4:104–112.

Birchfield, W. 1962. Host-parasite relations of *Rotylenchulus reniformis* on *Gossypium hirsutum.* Phytopathology 52:862–865.

Birchfield, W., and A. A. Antonopoulos. 1976. Scanning electron microscopic observations of *Duboscqia penetrans* parasitizing root-knot larvae. J. Nematol. 8:272–273.

Bird, A. F. 1959. The attractiveness of roots to the plant parasitic nematodes *Meloidogyne javanica* and *M. hapla.* Nematologica 4:322–335.

Bird, A. F. 1960. The effect of some single element deficiencies on the growth of *Meloidogyne javanica.* Nematologica 5:78–85.

Bird, A. F. 1962. Orientation of the larvae of *Meloidogyne javanica* relative to roots. Nematologica 8:275–287.

Bird, A. F. 1969. The influence of tobacco ring spot virus and tobacco mosaic virus on the growth of *Meloidogyne javanica.* Nematologica 15:201–209.

Bird, A. F. 1970. The effect of nitrogen deficiency on the growth of *Meloidogyne javanica* at different population levels. Nematologica 16:13–21.

Bird, A. F. 1972. Influence of temperature on embryogenesis in *Meloidogyne javanica.* J. Nematol. 4:206–213.

Bird, A. F. 1974. Suppression of embryogenesis and hatching in *Meloidogyne javanica* by thermal stress. J. Nematol. 6:95–99.

Bird, G. W., and W. R. Jenkins. 1964. Occurrence, parasitism, and pathogenicity of nematodes associated with cranberry. Phytopathology 54:677–680.

Bird, G. W., and W. R. Jenkins. 1965. Effect of cranberry bog flooding and low dissolved oxygen concentrations on nematode populations. Plant Dis. Rep. 49:517–518.

Bird, G. W., and W. F. Mai. 1967. Factors influencing population densities of *Trichodorus christiei*. Phytopathology 57:1368–1371.

Bird, G. W., S. M. McCarter, and R. W. Roncadori. 1971. Role of nematodes and soil-borne fungi in cotton stunt. J. Nematol. 3:17–22.

Bird, G. W., J. L. Crawford, and N. E. McGlohon. 1973. Distribution, frequency of occurrence, and population dynamics of *Rotylenchulus reniformis* in Georgia. Plant Dis. Rep. 57:399–401.

Bird, G. W., O. L. Brooks, C. E. Perry, J. C. Futral, T. D. Canerday, and F. C. Boswell. 1974. Influence of subsoiling and soil fumigation on the cotton stunt disease complex, *Hoplolaimus columbus* and *Meloidogyne incognita*. Plant Dis. Rep. 58:541–544.

Blair, G. P., and H. M. Darling. 1968. Red ring disease of the coconut palm, inoculation studies and histopathology. Nematologica 14:395–403.

Blake, C. D. 1961. Importance of osmotic potential as a component of the total potential of the soil water on the movement of nematodes. Nature 192:144–145.

Blake, C. D. 1962. Some observations on the orientation of *Ditylenchus dipsaci* and invasion of oat seedlings. Nematologica 8:177–192.

Bloom, J. R. 1963. Effect of temperature extremes on the wheat seed gall nematode, *Anguina tritici*. Plant Dis. Rep. 47:938–940.

Boosalis, M. G., and R. Mankau. 1965. Parasitism and predation of soil microorganisms. Pp. 374–391 *in* K. F. Baker and W. C. Snyder, Eds., Ecology of soil-borne plant pathogens. University of California Press, Berkeley, Calif.

Boughey, A. S. 1973. Ecology of populations. Macmillan, New York.

Brathwaite, C. W. D. 1974. Effect of crop sequence and fallow on populations of *Rotylenchulus reniformis* in fumigated and untreated soil. Plant Dis. Rep. 58:259–261.

Bray, J. R., and J. T. Curtis. 1957. An ordination of the upland forest communities of southern Wisconsin. Ecol. Monogr. 27:325–349.

Bridge, J., and N. G. M. Hague. 1974. The feeding behaviour of *Tylenchorhynchus* and *Merlinius* species and their effect on growth of perennial ryegrass. Nematologica 20:119–130.

Brock, T. D. 1966. Principles of microbial ecology. Prentice-Hall, Englewood Cliffs, N.J.

Brodie, B. B. 1976a. Effects of birds ingesting *Heterodera rostochiensis* cysts on viability of eggs and larvae. J. Nematol. 8:318–322.

Brodie, B. B. 1976b. Vertical distribution of three nematode species in relation to certain soil properties. J. Nematol. 8:243–247.

Brodie, B. B., and W. S. Murphy. 1975. Population dynamics of plant nematodes as affected by combinations of fallow and cropping sequence. J. Nematol. 7:91–92.

Brodie, B. B., J. M. Good, and W. E. Adams. 1969. Population dynamics of plant nematodes in cultivated soil: Effect of sod-based rotations in Cecil sandy loam. J. Nematol. 1:309–312.

Brodie, B. B., J. M. Good, and C. A. Jaworski. 1970a. Population dynamics of plant nematodes in cultivated soil: Effect of summer cover crops in old agricultural land. J. Nematol. 2:147–151.

Brodie, B. B., J. M. Good, and C. A. Jaworski. 1970b. Population dynamics of plant nematodes in cultivated soil: Effect of summer cover crops in newly cleared land. J. Nematol. 2:217–222.

Brodie, B. B., J. M. Good, and W. H. Marchant. 1970c. Population dynamics of plant nematodes in cultivated soil: Effect of sod-based rotations in Tifton sandy loam. J. Nematol. 2:135–138.

Brown, L. N. 1933. Flooding to control root-knot nematodes. J. Agric. Res. 47:883–888.

Brown, W. L., Jr. 1954. Collembola feeding upon nematodes. Ecology 35:421.

Brzeski, M. 1962a. Nematodes of peat-mosses of the Bialowieza Forest. Acta Zool. Cracov. 7:53–62.

Brzeski, M. 1962b. The nematodes of the peat mosses in the Kóscieliska Valley (The Western Tatra). Acta Zool. Cracov. 7:24–37.

Brzeski, M. 1963a. Further studies on nematodes (Nematoda) of the *Sphagnaceae* of the Tatra Mountains. Fragments Faun. 10:309–315.

Brzeski, M. 1963b. *Tylenchus ditissimus* sp. n., a new nematode from Poland (Nematoda, Tylenchidae). Bull. Acad. Polon. Sci. Ce II 11:537–540.

Brzeski, M. W. 1965. Nematodes associated with citrus trees infected by four viruses and comments about nematode distribution in Florida citrus groves. Plant Dis. Rep. 49:610–614.

Brzeski, M. W. 1969. Nematodes associated with cabbage in Poland. III. Host-parasite relations of *Heterodera schachti* Schmidt. Ekologia Polska. Seria A. 17:227–240.

Brzeski, M. W. 1970a. Plant parasitic nematodes associated with carrot in Poland. Roczniki Nauk Rolnic. Seria E 1:93–102.

Brzeski, M. W. 1970b. The effect of temperature on development of *Heterodera schachtii* Schm. and *Tylenchorhynchus dubius* But. (Nematoda, Tylenchida). Roczniki Nauk Rolnic. Seria E 1:205–211.

Brzeski, M. W., and A. Dowe. 1969. Effect of pH on *Tylenchorhynchus dubius* (Nematoda, Tylenchidae). Nematologica 15:403–407.

Bütschli, O. 1878. Beiträge zur Kenntniss der Flagellaten und einiger verwandten organismen. Zeitschift Wiss. Zool. 30:205–281.

Burns, N. C. 1971. Soil pH effects on nematode populations associated with soybeans. J. Nematol. 3:238–245.

Byars, L. P. 1919. The eelworm disease of wheat and its control. Farmers' Bull. 1041. 10 p.

Byars, L. P. 1920. The nematode disease of wheat caused by *Tylenchus tritici*. U.S. Dept. Agric. Bull. 842. 40 p.

Cairns, E. J. 1953. A culture-reared, plant parasitic nematode suitable for teaching and research. Phytopathology 43:105–106.

Canning, E. U. 1973. Protozoal parasites as agents for biological control of plant-parasitic nematodes. Nematologica 19: 342–348.

Carne, W. M. 1926. Ear cockle (*Tylenchus tritici*) and a bacterial disease (*Pseudomonas tritici*) of wheat. J. West Austr. Agric. Dept. (Ser. 2) 3:508–512.

Carter, W. W. 1975. Effects of soil texture on the interaction between *Rhizoctonia solani* and *Meloidogyne incognita* on cotton seedlings. J. Nematol. 7:234–236.

Carter, W. W., and S. Nieto, Jr. 1975. Population development of *Meloidogyne incognita* as influenced by crop rotation and fallow. Plant Dis. Rep. 59:402–403.

Castaner, D. 1963. Nematode populations in corn plots receiving different soil amendments. Proc. Iowa Acad. Sci. 70:107–113.

Castaner, D. 1966. The relationship of numbers of *Helicotylenchus microlobus* in nitrogen soil amendments. Iowa State J. Sci. 41:125–135.

Cauquil, J., and R. L. Shepherd. 1970. Effect of root-knot nematode–fungi combinations on cotton seedling disease. Phytopathology 60:448–451.

Caveness, F. E. 1958. The incidence of *Heterodera schachtii* soil population densities in various soil types. J. Amer. Soc. Sugar Beet Technol. 10:177–180.

Chang, H. Y., and D. J. Raski. 1972. *Hemicriconemoides chitwoodi* on grapevines. Plant Dis. Rep. 56:1028–1030.

Chantanao, A., and H. J. Jensen. 1969. Saprozoic nematodes as carriers and disseminators of plant pathogenic bacteria. J. Nematol. 1:216–218.

Chapman, R. A. 1959. Development of *Pratylenchus penetrans* and *Tylenchorhynchus martini* on red clover and alfalfa. Phytopathology 49:357–359.

Chapman, R. A. 1963. Population development of the plant parasitic nematode *Scutellonema brachyrum* on red clover. Proc. Helminthol. Soc. Wash. 30:169–173.

Chapman, R. N. 1928. The quantitative analysis of environmental factors. Ecology 1:111–122.

Cheo, C. C. 1946. A note on the relation of nematodes (*Tylenchus tritici*) to the development of the bacterial disease of wheat caused by *Bacterium tritici*. Ann. Appl. Biol. 33:446–449.

Chitwood, B. G. 1949. "Root-knot nematodes"—Part I. A revision of the genus *Meloidogyne* Goeldi, 1887. Proc. Helminthol. Soc. Wash. 16:90–104.

Chitwood, B. G. 1951a. North American marine nematodes. Tex. J. Sci. 3:617–672.

Chitwood, B. G. 1951b. The golden nematode of potatoes. U.S. Dept. Agric. Circ. 875. 48 p.

Chitwood, B. G. 1957. Two new species of the genus *Criconema* Hofmänner and Menzel. 1914. Proc. Helminthol. Soc. Wash. 24:57–61.

Chitwood, B. G., and M. B. Chitwood. 1950. An introduction to nematology. Monumental Printing, Baltimore, Md.

Christie, J. R. 1933. Further notes on the nematodes associated with the soreshin of cotton. Plant Dis. Rep. 17:10–12.

Christie, J. R. 1938. Two nematodes associated with decaying citrus fruit. Proc. Helminthol. Soc. Wash. 5:29–33.

Christie, J. R. 1952. Some new nematode species of critical importance to Florida growers. Proc. Soil Sci. Soc. Fla. 12:30–39.

Christie, J. R. 1959. Plant nematodes—their bionomics and control. University of Florida Press, Gainesville.

Christie, J. R. 1960. Biological control—predaceous nematodes. Pp. 466–468 *in* J. N. Sasser and W. R. Jenkins, Eds., *Nematology*. University North Carolina Press, Chapel Hill, N.C.

Christie, J. R., and C. H. Arndt. 1936. Feeding habits of the nematodes *Aphelenchoides parietinus* and *Aphelenchus avenae*. Phytopathology 26:698–701.

Christie, J. R., and W. Birchfield. 1958. Scribner's lesion nematode, a destructive parasite of amaryllis. Plant Dis. Rep. 42:873–875.

Christie, J. R., and L. Crossman. 1936. Notes on the strawberry strains of the bud and leaf nematode, *Aphelenchoides fragariae*, I. Proc. Helminthol. Soc. Wash. 3:69–72.

Churchill, R. C., Jr., and J. L. Ruehle. 1971. Occurrence, parasitism, and pathogenicity of nematodes associated with sycamore (*Platanus occidentalis* L.). J. Nematol. 3:189–196.

Cobb, N. A. 1914. The North American free-living fresh-water nematodes. Trans. Amer. Microsc. Soc. 33:69–134.

Cobb, N. A. 1915. Nematodes and their relationships. U.S. Dept. Agric. Yearbook for 1914:457–490.

Cobb, N. A. 1920. Transference of nematodes (Mononchs) from place to place for economic purposes. Science 51:640–641.

Cobb, N. A. 1924. Interesting new genus of nemas inhabiting nests of tropical ants. J. Parasitol. 10:209–210.

Cobb, N. A. 1929. Characteristics of carnivorous free-living nemas, including predators. J. Parasitol. 15:284–285.

Cockerell, T. D. A. 1934. 'Mimicry' among insects. Nature 133:329–330.

Cohn, E. 1965. On the feeding and histopathology of the citrus nematode. Nematologica. 11:47–54.

Cohn, E. 1969. The occurrence and distribution of species of *Xiphinema* and *Longidorus* in Israel. Nematologica 15:179–192.

Cohn, E. 1970. Observations on the feeding and symptomatology of *Xiphinema* and *Longidorus* on selected host roots. J. Nematol. 2:167–173.

Cohn, E., and M. Mordechai. 1969. Investigations on the life cycles and host preference of some species of *Xiphinema* and *Longidorus* under controlled conditions. Nematologica 15:295–302.

Cohn, E., and M. Mordechai. 1970. The influence of some environmental and cultural conditions on rearing populations of *Xiphinema* and *Longidorus*. Nematologica 16:85–93.

Collins, D. D. 1966. The occurrence of seed-gall nematodes in a native population of western wheatgrass (*Agropyron smithii*) in Montana. Plant Dis. Rep. 50:45.

Collis-George, N., and C. D. Blake. 1959. The influence of the soil moisture regime on the expulsion of the larval mass of the nematode *Anguina agrostis* from galls. Austral. J. Biol. Sci. 12:247–256.

Comandon, J., and P. de Fonbrune. 1938. Recherches expérimentales sur les champignons prédateurs de nématodes du sol. Compt. Rend. Soc. Biol. Paris 129:619–625.

Conroy, J. J., R. J. Green, Jr., and J. M. Ferris. 1972. Interaction of *Verticillium albo-atrum* and the root lesion nematode, *Pratylenchus penetrans*, in tomato roots at controlled inoculum densities. Phytopathology 62:362–366.

Cooke, R. C. 1962a. Behaviour of nematode-trapping fungi during decomposition of organic matter in the soil. Trans. Brit. Mycol. Soc. 45:314–320.

Cooke, R. C. 1962b. The ecology of nematode-trapping fungi in the soil. Ann. Appl. Biol. 50:507–513.

Cooke, R. C. 1963. Succession of nematophagous fungi during the decomposition of organic matter in the soil. Nature 197:205.

Cooke. R. C. 1964. Ecological characteristics of nematode-trapping Hyphomycetes. II. Germination of conidia in soil. Ann. Appl. Biol. 54:375–379.

Cooke, R. C. 1968. Relationships between nematode-destroying fungi and soil-borne phytonematodes. Phytopathology 58:909–913.

Cooper, A. F., Jr., S. D. Van Gundy, and L. H. Stolzy. 1970. Nematode reproduction in environments of fluctuating aeration. J. Nematol. 2:182–188.

Corbett, D. C. M., and R. M. Webb. 1970. Plant and soil nematode population changes in wheat grown continuously in ploughed and in unploughed soil. Ann. Appl. Biol. 65:327–335.

Coscarelli, W., and D. Pramer. 1962. Nutrition and growth of *Arthrobotrys conoides*. J. Bacteriol. 84:60–64.

Cotten, J. 1973. Feeding behaviour and reproduction of *Xiphinema index* on some herbaceous test plants. Nematologica 19:516–520.

Couch, J. N. 1937. The formation and operation of the traps in the nematode-catching fungus *Dactylella bembicodes* Drechsler. J. Elisha Mitchell Sci. Soc. 53:301–309.

Courtney, W. D., and H. B. Howell. 1952. Investigations on the bent grass nematode, *Anguina agrostis* (Steinbuch 1799) Filipjev 1936. Plant Dis. Rep. 36:75–83.

Cralley, E. M. 1957. The effect of seeding methods on the severity of white tip of rice. Phytopathology 47:7. (Abstr.)

Croll. N. A. 1967. Acclimatization in the eccritic thermal response of *Ditylenchus dipsaci*. Nematologica 13:385–389.

Curtis, J. T. 1959. The vegetation of Wisconsin. An ordination of plant communities. University of Wisconsin Press, Madison.

Darrah, W. C. 1960. Principles of paleobotany. Ronald, New York.

Darwin, C. 1859. The origin of species. Murray, London.

Das, S., and D. J. Raski. 1969. Effect of grapevine fanleaf virus on the reproduction and survival of its nematode vector, *Xiphinema index* Thorne & Allen. J. Nematology 1:107–110.

Das, V. M. 1964. *Hexatylus mulveyi* n. sp. and *Deladenus durus* (Cobb, 1922) Thorne, 1941 (Nematoda: Neotylenchidae) from the Canadian Arctic. Can. J. Zool. 42:649–653.

Dasgupta, D. R., and D. J. Raski. 1968. The biology of *Rotylenchulus parvus*. Nematologica 14:429–440.

Daulton, R. A. C., and C. J. Nusbaum. 1961. The effect of soil temperature on the survival of the root-knot nematodes *Meloidogyne javanica* and *M. hapla*. Nematologica 6:280–294.

Davaine, C. 1857. Recherches sur l'anguillule du blé niellé considérée au point de vue de l'histoire naturelle et de l'agriculture. Compt. Rend. Soc. Biol. (1856) (Memoires) 8:201–271.

Davide, R. G., and A. C. Triantaphyllou. 1967a. Influence of the environment on development and sex differentiation of root-knot nematodes. I. Effect of infection density, age of host plant and soil temperature. Nematologica 13:102–110.

Davide, R. G., and A. C. Triantaphyllou. 1967b. Influence of the environment of development and sex differentiation of root-knot nematodes. II. Effect of host nutrition. Nematologica 13:111–117.

Davide, R. G., and A. C. Triantaphyllou. 1968. Influence of the environment on development and sex differentiation of root-knot nematodes. III. Effect of foliar application of maleic hydrazide. Nematologica 14:37–46.

Davidson, J., and G. J. Curtis. 1973. An Iron Age site on the land of the Plant Breeding Institute, Trumpington. Proc. Camb. Ant. Soc. 64:1–14.

Davis, R. A., and W. R. Jenkins. 1960. Nematodes associated with roses and the root injury caused by *Meloidogyne hapla* Chitwood, 1949, *Xiphinema diversicaudatum* (Micoletzky, 1927), Thorne, 1939, and *Helicotylenchus nanus* Steiner, 1945. Univ. Md. Agr. Exp. Sta. Bull. A-106. 16 p.

Deubert, K. H., and R. A. Rohde. 1971. Nematode enzymes. Pp. 73–90 *in* B. M. Zuckerman, W. F. Mai, and R. A. Rohde, Eds., Plant parasitic nematodes. Vol. II. Academic, New York.

D'Herde, J., and J. van den Brande. 1964. Distribution of *Xiphinema* and *Longidorus* spp. in strawberry fields in Belgium and a method for their quantitative extraction. Nematologica 10:454–458.

Dickerson, O. J., H. M. Darling, and G. D. Griffin. 1964. Pathogenicity and population trends of *Pratylenchus penetrans* on potato and corn. Phytopathology 54:317–322.

Di Edwardo, A. A. 1961. Seasonal population variations of *Pratylenchus penetrans* in and about strawberry roots. Plant Dis. Rep. 45:67–71.

Ditlevsen, H. 1927. Free-living nematodes from Greenland, land and freshwater. Meddelelser Gronland 23 (Suppl.):157–198.

Dollfus, R. P. 1946. Parasites (animaux et vegetaux) des Helminthes. Encyclopedie Biol. 27. Lechevalier, Paris.

Dolliver, J. S. 1961. Population levels of *Pratylenchus penetrans* as influenced by treatments affecting dry weight of Wando pea plants. Phytopathology 51:364–367.

Donaldson, F. S., Jr. 1967. *Meloidogyne javanica* infesting *Pinus elliottii* seedlings in Florida. Plant Dis. Rep. 51:455–456.

Doncaster, C. C. 1953. A study of the host-parasite relationships. The potato-root eelworm (*Heterodera rostochiensis*) in black nightshade (*Solanum nigrum*) and tomato. J. Helminthol. 27:1–8.

Doncaster, C. C., and D. J. Hooper. 1961. Nematodes attacked by protozoa and tardigrades. Nematologica 6:333–335.

Dropkin, V. H., G. C. Martin, and R. W. Johnson. 1958. Effect of osmotic concentration on hatching of some plant parasitic nematodes. Nematologica 3:115–126.

DuCharme, E. P. 1955. Sub-soil drainage as a factor in the spread of the burrowing nematode. Proc. Fla. State Hort. Soc. 68:29–31.

DuCharme, E. P. 1959. Morphogenesis and histopathology of lesions induced on citrus roots by *Radopholus similis*. Phytopathology 49:388–395.

DuCharme, E. P. 1967. Annual population periodicity of *Radopholus similis* in Florida citrus groves. Plant Dis. Rep. 51:1031–1034.

DuCharme, E. P., and W. C. Price. 1966. Dynamics of multiplication of *Radopholus similis*. Nematologica 12:113–121.

Duddington, C. L. 1957. The friendly fungi. Faber & Faber, London.

Duddington, C. L., and C. M. G. Duthoit. 1960. Green manuring and cereal root eelworm. Plant Pathol. 9:7–9.

Duddington, C. L., F. G. W. Jones, and F. Moriarty. 1956. The effect of predacious fungus and organic matter upon the soil population of beet eelworm, *Heterodera schachtii* Schm. Nematologica 1:344–348.

Duddington, C. L., C. O. R. Everard, and C. M. G. Duthoit. 1961. Effect of green manuring and a predacious fungus on cereal root eelworm in oats. Plant Pathol. 10:108–109.

Duggan, J. J. 1963. Relationship between intensity of cereal root eelworm (*Heterodera avenae* Wollenweber 1924) infestation and pH value of soil. Irish J. Agric. Res. 2:105–110.

El-Goorani, M. A., M. K. A-E-Dahab, and F. F. Mehiar. 1974. Interaction between root knot nematode and *Pseudomonas marginata* on gladiolus corms. Phytopathology 64:271–272.

Eliava, I. Y. 1966. Nematode fauna of mosses in the Georgian SSR. Pp. 5–10 *in* Materials on the fauna of the Georgian SSR. Izdatelstra Akad. Nauk Gruz. SSR. No. 1.

Ellenby, C. 1946. Ecology of the eelworm cyst. Nature 157:451–452.

Ellenby, C. 1954. Environmental determination of the sex ratio of a plant parasitic nematode. Nature 174:1016–1017.

Ellenby, C. 1955. The seasonal response of the potato-root eelworm *Heterodera rostochiensis* Wollenweber: Emergence of larvae throughout the year from cysts exposed to different temperature cycles. Ann. Appl. Biol. 43:1–11.

Ellenby, C. 1968. The survival of desiccated larvae of *Heterodera rostochiensis* and *H. schachtii*. Nematologica 14:544–548.

Elmiligy, I. A. 1968. Root-knot nematode infectivity and host response in relation to soil types. Meded. Rijk. Landbouweten. Gent. 33:1633–1641.

Elmiligy, I. A. 1971. Recovery and survival of some plant-parasitic nematodes as influenced by temperature, storage time and extraction technique. Meded. Rijk. Landbouweten. Gent. 36:1333–1339.

Elmiligy, I. A., and D. C. Norton. 1973. Survival and reproduction of some nematodes as affected by muck and organic acids. J. Nematol. 5:50–54.

El-Sherif, M., and W. F. Mai. 1969. Thermotactic response of some plant parasitic nematodes. J. Nematol. 1:43–48.

Endo, B. Y. 1959. Responses of root-lesion nematodes, *Pratylenchus brachyurus* and *P. zeae*, to various plants and soil types. Phytopathology 49:417–421.

Eno, C. F., W. G. Blue, and J. M. Good, Jr. 1955. The effect of anhydrous ammonia on nematodes, fungi, bacteria, and nitrification in some Florida soils. Proc. Soil Sci. Soc. Amer. 19:55–58.

Epps, J. M. 1969. Survival of the soybean cyst nematode in seed stocks. Plant Dis. Rep. 53:403–405.

Epps, J. M. 1971. Recovery of soybean cyst nematodes (*Heterodera glycines*) from the digestive tract of blackbirds. J. Nematol. 3:417–419.

Esser, R. P. 1963. Nematode interactions in plates of non-sterile water agar. Proc. Soil and Crop Sci. Soc. Fla. 23:121–138.

Esser, R. P. 1964. Plant parasitic nematodes associated with Sabal palmetto on virgin land in the Florida Everglades. Plant Dis. Rep. 48:533.

Esser, R. P., and W. H. Ridings. 1974. Pathogenicity of selected nematodes by *Catenaria anguillulae*. Proc. Soil and Crop Sci. Soc. Fla. 33:60–64.

Esser, R. P., and E. K. Sobers. 1964. Natural enemies of nematodes. Proc. Soil and Crop Sci. Soc. Fla. 24:326–353.

Estores, R. A., and T. A. Chen. 1972. Interactions of *Pratylenchus penetrans* and *Meloidogyne incognita* as coinhabitants in tomato. J. Nematol. 4:170–174.

Evans, A. A. F. 1970. Mass culture of mycophagous nematodes. J. Nematol. 2:99–100.

Evans, A. A. F., and J. M. Fisher. 1969. Development and structure of populations of *Ditylenchus myceliophagus* as affected by temperature. Nematologica 15:395–402.

Evans, A. A. F., and J. M. Fisher. 1970. Some factors affecting the number and size of nematodes in populations of *Aphelenchus avenae*. Nematologica 16:295–304.

Evans, K. 1970. Longevity of males and fertilisation of females of *Heterodera rostochiensis*. Nematologica 16:369–374.

Fassuliotis, G., and G. J. Rau. 1966. Observations on the embryogeny and histopathology of *Hypsoperine spartinae* on smooth cordgrass roots *Spartina alterniflora*. Nematologica 19:90. (Abstr.)

Fassuliotis, G., and G. J. Rau. 1969. The relationship of *Meloidogyne incognita acrita* to the incidence of cabbage yellows. J. Nematol. 1:219–222.

Faulkner, L. R. 1964. Pathogenicity and population dynamics of *Paratylenchus hamatus* on *Mentha* spp. Phytopathology 54:344–348.

Faulkner, L. R., and W. J. Bolander. 1966. Occurrence of large nematode populations in irrigation canals of South Central Washington. Nematologica 12:591–600.

Faulkner, L. R., and W. J. Bolander. 1969. Interaction of *Verticillium dahliae* and

Pratylenchus minyus in Verticillium wilt of peppermint: Effect of soil temperature. Phytopathology 59:868–870.

Faulkner, L. R., and W. J. Bolander. 1970a. Acquisition and distribution of nematodes in irrigation waterways of the Columbia Basin in eastern Washington. J. Nematol. 2:362–367.

Faulkner, L. R., and W. J. Bolander. 1970b. Agriculturally-polluted irrigation water as a source of plant-parasitic nematode infestation. J. Nematol. 2:368–374.

Faulkner, L. R., and H. M. Darling. 1961. Pathological histology, hosts, and culture of the potato rot nematode. Phytopathology 51:778–786.

Faulkner, L. R., and C. B. Skotland. 1965. Interactions of *Verticillium dahliae* and *Pratylenchus minyus* in Verticillium wilt of peppermint. Phytopathology 55:583–586.

Faulkner, L. R., W. J. Bolander, and C. B. Skotland. 1970. Interaction of *Verticillium dahliae* and *Pratylenchus minyus* in Verticillium wilt of peppermint: Influence of the nematode as determined by a double root technique. Phytopathology 60:100–103.

Fawcett, H. S. 1931. The importance of investigations on the effects of known mixtures of microorganisms. Phytopathology 21:545–550.

Feder, W. A., and J. Feldmesser. 1961. The spreading decline complex: The separate and combined effects of *Fusarium* spp. and *Radopholus similis* on the growth of Duncan grapefruit seedlings in the greenhouse. Phytopathology 51:724–726.

Feder, W. A., C. O. R. Everard, and L. M. O. Wootton. 1963. Sensitivity of several species of the nematophagous fungus *Dactylella* to a morphogenic substance derived from free-living nematodes. Nematologica 9:49–54.

Feldmesser, J., R. C. Cetas, G. R. Grimm, R. V. Rebois, and R. W. Whidden. 1960. Movement of *Radopholus similis* into rough lemon feeder roots and in soil and its relation to *Fusarium* in the roots. Phytopathology 50:635. (Abstr.)

Fenwick, D. W. 1951. Investigations on the emergence of larvae from cysts of the potato-root eelworm, *Heterodera rostochiensis*. 4. Physical conditions and their influence on larval emergence in the laboratory. J. Helminthol. 25:37–48.

Ferris, H. 1976. Development of a computer-simulation model for a plant-nematode system. J. Nematol. 8:255–263.

Ferris, H., and M. V. McKenry. 1974. Seasonal fluctuations in the spatial distribution of nematode populations in a California vineyard. J. Nematol. 6:203–210.

Ferris, J. M. 1967. Factors influencing the population fluctuation of *Pratylenchus penetrans* in soils of high organic content. I. Effect of soil fumigants and different crop plants. J. Econ. Entomol. 60:1708–1714.

Ferris, J. M. 1970. Soil temperature effects on onion seedling injury by *Pratylenchus penetrans*. J. Nematol. 2:248–251.

Ferris, V. R. 1961. A new species of *Pratylenchus* (Nemata-Tylenchida) from roots of soybeans. Proc. Helminthol. Soc. Wash. 28:109–111.

Ferris, V. R. 1963. *Tylenchorhynchus silvaticus* n. sp. and *Tylenchorhynchus agri* n. sp. (Nematoda: Tylenchida). Proc. Helminthol. Soc. Wash. 30:165–168.

Ferris, V. R., and R. L. Bernard. 1967. Population dynamics of nematodes in fields planted to soybeans and crops grown in rotation with soybeans. I. The genus *Pratylenchus* (Nemata: Tylenchida). J. Econ. Entomol. 60:405–410.

Ferris, V. R., and R. L. Bernard. 1971. Crop rotation effects on population densities of ectoparasitic nematodes. J. Nematol. 3:119–122.

Ferris, V. R., and J. M. Ferris. 1967. Morphological variant of *Tetylenchus joctus* Thorne (Nemata: Tylenchida) associated with cultivated blueberries in Indiana. Proc. Helminthol. Soc. Wash. 34:30–32.

Ferris, V. R., C. G. Goseco, and J. M. Ferris. 1976. Biogeography of free-living soil nematodes from the perspective of plate tectonics. Science 193:508–510.

Fidler, J. H., and W. J. Bevan. 1963. Some soil factors influencing the density of cereal root eelworm (*Heterodera avenae* Woll.) populations and their damage to the oat crop. Nematologica 9:412–420.

Fielding, M. J. 1951. Observations on the length of dormancy in certain plant infecting nematodes. Proc. Helminthol. Soc. Wash. 18:110–112.

Fisher, J. M. 1969. Investigations on fecundity of *Aphelenchus avenae*. Nematologica 15:22–28.

Fisher, K. D. 1968. Population patterns of nematodes in Lake Champlain. Nematologica 14:7 (Abstr.)

Flores, H., and R. A. Chapman. 1968. Population development of *Xiphinema americanum* in relation to its role as a vector of tobacco ringspot virus. Phytopathology 58:814–817.

Ford, H. W. 1953. Effect of spreading decline disease on the distribution of feeder roots of orange and grapefruit trees on rough lemon rootstock. Amer. Soc. Hort. Sci. 61:68–72.

Foster, J. W. 1949. Chemical activities of fungi. Academic, New York.

Franklin, M. T. 1955. A redescription of *Aphelenchoides parietinus* (Bastian, 1865) Steiner, 1932. J. Helminthol. 29:65–76.

Freckman, D. W., and R. A. Chapman. 1972. Infection of red clover seedlings by *Heterodera trifolii* Goffart and *Pratylenchus penetrans* (Cobb). J. Nematol. 4:23–28.

Freckman, D. W., R. Mankau, and S. A. Sher. 1974. Population dynamics of nematodes associated with dominant desert shrubs. J. Nematol. 6:140. (Abstr.)

Freckman, D. W., R. Mankau, and H. Ferris. 1975. Nematode community structure in desert soils: Nematode recovery. J. Nematol. 7:343–346.

French, N., and R. M. Barraclough. 1961. Observations on the reproduction of *Aphelenchoides ritzemabosi* (Schwartz). Nematologica 6:89–94.

Frey, D. G. 1965. Other invertebrates—an essay in biogeography. Pp. 613–631 in H. E. Wright, Jr. and D. G. Frey, Eds, The Quaternary of the United States. Princeton University Press, Princeton, N.J.

Fushtey, S. G., and P. W. Johnson. 1966. The biology of the oat cyst nematode, *Heterodera avenae* in Canada. I. The effect of temperature on the hatchability of cysts and emergence of larvae. Nematologica 12:313–320.

Gadea, E. 1964. Sobre la nematofauna muscicola de la islas Medas. Publnes. Inst. Biol. Apl., Barcelona 36:29–38.

Gadea, E. 1965. Sobre la nematofauna brioedafica de la islas Canarias. Publnes. Inst. Biol. Apl., Barcelona 38:79–91.

Gause, G. F. 1934. The struggle for existence. Williams and Wilkins, Baltimore.

Gay, C. M., and G. W. Bird. 1973. Influence of concomitant *Pratylenchus brachyurus* and *Meloidogyne* spp. on root penetration and population dynamics. J. Nematol. 5:212–217.

Geraert, E. 1967. Results of a study on the oecology of plant-parasitic and freeliving soil-nematodes. Ann. Soc. Roy. Zool. Belg. 97:59–64.

Gillespie, W. H., and R. E. Adams. 1962. An awl nematode, *Dolichodorus silvestris* n. sp. from West Virginia. Nematologica 8:93–98.

Girard, A. 1887. Sur le développement des nématodes de la betterave, pendant les années 1885 et 86, et sur leurs modes de propagation. Compt. Rend. de l'Acad. Sci. Bd. 104. S:522–524.

Girard, D. H. 1969. List of intercepted plant pests, 1968. United States Dept. Agr. ARS 82-6-3. 86 p.

Godfrey, G. H. 1940. Ecological specialization in the stem- and bulb-infesting nematode, *Ditylenchus dipsaci* var. *amsinckiae*. Phytopathology 30:41–53.

Godfrey, G. H. 1947. A practical control for nematodes. Proc. 2nd Ann. Lower Rio Grande Valley Citrus and Vegetable Institute, 1947:143–149.

Godfrey, G. H., and H. R. Hagan. 1933. Influence of soil hydrogen-ion concentration on infection by *Heterodera radicicola* (Greef) Muller. Soil Sci. 35:175–184.

Golden, A. M. 1956. Taxonomy of the spiral nematodes (*Rotylenchus* and *Helicotylenchus*), and the developmental stages and host–parasite relationships of *R. buxophilus* n. sp., attacking boxwood. Univ. Md. Agric. Exp. Sta. Bull. A-85. 28 p.

Golden, A. M. 1957. Occurrence of *Radopholus gracilus* (Nematoda : Tylenchidae) in the United States. Plant Dis. Rep. 41:91.

Golden, A. M. 1958. *Dolichodorus similis*, (Dolichodorinae), a new species of plant nematode. Proc. Helminthol. Soc. Wash. 25:17–20.

Golden, A. M., and O. J. Dickerson. 1973. *Heterodera longicolla*, n. sp. (Nematoda : Heteroderidae) from buffalo-grass (*Buchloë dactyloides*) in Kansas. J. Nematol. 5:150–154.

Golden, A. M., and H. J. Jensen. 1974. *Nacobbodera chitwoodi* n. gen., n. sp., (Nacobbidae : Nematoda) on Douglas fir in Oregon. J. Nematol. 6:30–37.

Golden, J. K., and S. D. Van Gundy. 1975. A disease complex of okra and tomato involving the nematode, *Meloidogyne incognita,* and the soil-inhabiting fungus, *Rhizoctonia solani*. Phytopathology 65:265–273.

Good, J. M., L. W. Boyle, and R. O. Hammons. 1958. Studies of *Pratylenchus brachyurus* on peanuts. Phytopathology 48:530–535.

Good, J. M., W. S. Murphy, and B. B. Brodie. 1973. Population dynamics of plant nematodes in cultivated soils: Length of rotation in newly cleared and old agricultural land. J. Nematol. 5:117–122.

Goodey, J. B. 1958. *Ditylenchus myceliophagus* n. sp. (Nematoda : Tylenchidae). Nematologica 3:91–96.

Goodey, T. 1933. *Anguillulina graminophila* n. sp., a nematode causing galls on the leaves of fine bent-grass. J. Helminthol. 11:45–56.

Goodey, T., and M. J. Triffitt. 1927. On the presence of flagellates in the intestine of the nematode *Diplogaster longicauda.* Protozoology (Supple. to J. Helminthol.) 3:47–58.

Goto, S., and J. W. Gibler. 1951. A leaf gall forming nematode on *Calamagrostis canadensis* (Michx.) Beauv. Plant Dis. Rep. 35:215–216.

Gotoh, A. 1970. The root-lesion nematodes found in uncultivated soils, mainly in natural grasslands in Japan. Proc. Assoc. Plant Prot. Hyushu Agric. Expt. Sta. 16:34–37.

Gowen, S. R. 1970. Observations on the fecundity and longevity of *Tylenchus emarginatus* on Sitka spruce seedlings at different temperatures. Nematologica 16:267–272.

Grandison, G. S., and H. R. Wallace. 1974. The distribution and abundance of *Pratylenchus thornei* in fields of strawberry clover (*Trifolium fragiferum*). Nematologica 20:283–290.

Grant, C. L., W. Coscarelli, and D. Pramer. 1962. Statistical measurement of biotin, thiamine, and zinc concentrations required for maximal growth of *Arthrobotrys conoides*. Appl. Microbial. 10:413–417.

Green, C. D. 1971. Mating and host finding behavior of plant nematodes. Pp. 247–266 *in* B. M. Zuckerman, W. F. Mai, and R. A. Rohde, Eds., Plant parasitic nematodes. Vol. II. Academic, New York.

Green, C. D., D. N. Greet, and F. G. W. Jones. 1970. The influence of multiple mating on the reproduction and genetics of *Heterodera rostochiensis* and *H. schachtii*. Nematologica 16:309–326.

Griffin, G. D. 1968. The pathogenicity of *Ditylenchus dipsaci* to alfalfa and its relationship of temperature to plant infection and susceptibility. Phytopathology 58:929–932.

Griffin, G. D. 1969. Effects of temperature on *Meloidogyne hapla* in alfalfa. Phytopathology 59:599–602.

Griffin, G. D. 1974. Effect of acclimation temperature on infection of alfalfa by *Ditylenchus dipsaci*. J. Nematol. 6:57–59.

Griffin, G. D., and K. R. Barker. 1966. Effects of soil temperature and moisture on the survival and activity of *Xiphinema americanum*. Proc. Helminthol. Soc. Wash. 33:126–130.

Griffin, G. D., and H. M. Darling. 1964. An ecological study of *Xiphinema americanum* Cobb in an ornamental spruce nursery. Nematologica 10:471–479.

Griffin, G. D., and O. J. Hunt. 1972. Effects of temperature and inoculation timing on the *Meloidogyne hapla*/*Corynebacterium insidiosum* complex in alfalfa. J. Nematol. 4:70–71.

Griffin, G. D., and E. C. Jorgenson. 1969, Life cycle and reproduction of *Meloidogyne hapla* on potato. Plant Dis. Rep. 53:259–261.

Griffin, G. D., J. L. Anderson, and E. C. Jorgenson. 1968. Interaction of *Meloidogyne hapla* and *Agrobacterium tumefaciens* in relation to raspberry cultivars. Plant Dis. Rep. 52:492–493.

Grisham, M. P., J. L. Dale, and R. D. Riggs. 1974. *Meloidogyne graminis* and *Meloidogyne* spp. on *Zoysia*; infection, reproduction, disease development, and control. Phytopathology 64:1485–1489.

Grundbacher, F. J., and E. H. Stanford. 1962. Effect of temperature on resistance of alfalfa to the stem nematode (*Ditylenchus dipsaci*). Phytopathology 52:791–794.

Gupta, P., and G. Swarup. 1972. Ear-cockle and yellow ear-rot diseases of wheat. II. Nematode bacterial association. Nematologica 18:320–324.

Habicht, W. A., Jr. 1975. The nematicidal effects of varied rates of raw and composted sewage sludge as soil organic amendments on a root-knot nematode. Plant Dis. Rep. 59:631–634.

Hagge, A. H. 1969. Soybean-cyst-nematode damage as associated with intensive culture of soybeans in southeastern Missouri. Pp. 41–48 *in* J. A. Browning, Ed., Disease consequences of intensive and extensive culture of field crops. Iowa Agric. and Home Econ. Exp. Sta. Spec. Rept. 64.

Hagley, E. A. C. 1963. The role of the palm weevil, *Rhynchophorus palmarum* as a vector of red ring disease of coconuts. I. Results of preliminary investigations. J. Econ. Entomol. 56:375–380.

Haglund, W. A., and T. H. King. 1961. Effect of parasitic nematodes on the severity of common root rot of canning peas. Nematologica 6:311–314.

Haglund, W. A., and D. R. Milne, Jr. 1973. Nematode dissemination in commercial mushroom houses. Phytopathology 63:1455–1458.

Hamblen, M. L., and D. A. Slack. 1959. Factors influencing the emergence of larvae from cysts of *Heterodera glycines* Ichinohe. Cyst development, condition, and variability. Phytopathology 49:317. (Abstr.)

Hams, A. F., and G. D. Wilkin. 1961. Observations on the use of predacious fungi for the control of *Heterodera* spp. Ann. Appl. Biol. 49:515–523.

Harrison, A. L., and P. A. Young. 1941. Effect of root-knot nematode on tomato wilt. Phytopathology 31:749–752.

Hawn, E. J. 1963. Transmission of bacterial wilt of alfalfa by *Ditylenchus dipsaci* Kühn. Nematologica 9:65–68.

Hawn, E. J., and M. R. Hanna. 1967. Influence of stem nematode infestation on bacterial wilt reaction and forage yield of alfalfa varieties. Can. J. Plant Sci. 47:203–208.

Hayes, W. A., and F. Blackburn. 1966. Studies on the nutrition of *Arthrobotrys oligospora* Fres. and *A. robusta* Dudd. II. The predaceous phase. Ann. Appl. Biol. 58:51–60.

Heald, C. M. 1975. Pathogenicity and histopathology of *Rotylenchulus reniformis* infecting cantaloup. J. Nematol. 7:149–152.

Heald, C. M., and G. W. Burton. 1968. Effect of organic and inorganic nitrogen on nematode populations on turf. Plant Dis. Rep. 52:46–48.

Heald, C. M., an M. D. Heilman. 1971. Interaction of *Rotylenchulus reniformis*, soil salinity, and cotton. J. Nematol. 3:179–182.

Hechler, H. C. 1963. Description, developmental biology, and feeding habits of *Seinura tenuicaudata* (De Man) J. B. Goodey, 1960 (Nematoda:Aphelenchoididae), a nematode predator. Proc. Helminthol. Soc. Wash. 30:182–195.

Hechler, H. C., and D. P. Taylor. 1965. Taxonomy of the genus *Seinura* (Nematoda:Aphelenchoididae), with descriptions of *S. celeris* n. sp. and *S. steineri* n. sp. Proc. Helminthol. Soc. Wash. 32:205–219.

Hechler, H. C., and D. P. Taylor. 1966. The life histories of *Seinura celeris, S. oliveirae, S. oxura,* and *S. steineri.* Proc. Helminthol. Soc. Wash. 33:71–83.

Hewitt, W. B., D. J. Raski, and A. C. Goheen. 1958. Nematode vector of soil-borne fanleaf virus of grapevines. Phytopathology 48:586–595.

Hirschmann, H. 1962. The life cycle of *Ditylenchus triformis* (Nematoda:Tylenchida) with emphasis on post-embryonic development. Proc. Helminthol. Soc. Wash. 29:30–43.

Hirschmann, H., and R. D. Riggs. 1969. *Heterodera betulae* n. sp. (Heteroderidae), a cyst-forming nematode from river birch. J. Nematol. 1:169–179.

Hirschmann, H., and J. N. Sasser. 1955. On the occurence of an intersexual form in *Ditylenchus triformis,* n. sp. (Nematoda:Tylenchidae). Proc. Helminthol. Soc. Wash. 22:115–123.

Hoff, J. K., and W. F. Mai. 1964. Influence of soil depth and sampling date on population levels of *Trichodorus christiei.* Phytopathology 54:246.

Hoffmann, J. K. 1974. Morphological variation in species of *Bakernema, Criconema,* and *Criconemoides* (Criconematidae:Nematoda). Iowa State J. Res. 49:137–153.

Hoffmann, J. K., and D. C. Norton. 1973. Distribution patterns of *Criconema* and *Criconemoides* in prairie, oak-hickory, hemlock-hardwood, boreal forest, and tundra communities. Abstract No. 1091, 2nd Int. Congr. Plant Pathol., Abstr. of papers, 1973. Amer. Phytopathol. Soc.

Hoffmann, J. K., and D. C. Norton. 1976. Distribution patterns of some Criconematinae in different forest associations. J. Nematol. 8:32–35.

Holdeman, Q. L. 1955. The present known distribution of the sting nematode, *Belonolaimus gracilis,* in the coastal plain of the southeastern United States. Plant Dis. Rep. 39:5–8.

Holdeman, Q. L. 1956. The effect of the tobacco stunt nematode on the incidence of Fusarium wilt in flue-cured tobacco. Phytopathology 46:129.

Holdeman, Q. L., and T. W. Graham. 1954. Effect of the sting nematode on expression of Fusarium wilt in cotton. Phytopathology 44:683–685.

Hollis, J. P. 1957. Cultural studies with *Dorylaimus ettersbergensis.* Phytopathology 47:468–473.

Hollis, J. P. 1958. Induced swarming of a nematode as a means of isolation. Nature 182:956–957.

Hollis, J. P. 1967. Nature of the nematode problem in Louisiana rice fields. Plant Dis. Rep. 51:167–169.

Hollis, J. P., and M. J. Fielding. 1958. Population behavior of plant parasitic nematodes in soil fumigation experiments. La. Agric. Exp. Sta. Bull. 515. 23 p.

Hollis, J. P., and R. Rodriguez-Kabana. 1966. Rapid kill of nematodes in flooded soil. Phytopathology 56:1015–1019.

Hopper, B. E. 1959. Three new species of the genus *Tylenchorhynchus* (Nematoda:Tylenchida). Nematologica 4:23–30.

Horne, C. W., and W. H. Thames, Jr. 1966. Notes on occurrence and distribution of *Heterodera punctata.* Plant Dis. Rep. 50:869–871.

Huang, C. S., and S. P. Huang. 1974. Dehydration and the survival of rice white tip nematode *Aphelenchoides besseyi.* Nematologica 20:9–18.

Huang, C. S., S. P. Huang, and L. H. Lin. 1972. The effect of temperature on development and generation periods of *Aphelenchoides besseyi*. Nematologica 18:432–438.

Hunt, O. J., G. D. Griffin, J. J. Murray, M. W. Pedersen, and R. N. Peaden. 1971. The effects of root knot nematodes on bacterial wilt of alfalfa. Phytopathology 61:256–259.

Hunt, P. G., G. C. Smart, Jr., and C. F. Eno. 1973. Sting nematode, *Belonolaimus longicaudatus*, immotility induced by extracts of composted municipal refuse. J. Nematol. 5:60–63.

Hussey, R. S. 1977. Effects of subsoiling and nematicides on *Hoplolaimus columbus* populations and cotton yield. J. Nematol. 9:83–86.

Hutchinson, M. T., and H. T. Streu. 1960. Tardigrades attacking nematodes. Nematologica 5:149.

Hutchinson, M. T., J. P. Reed, and D. Pramer. 1960. Observations on the effects of decaying vegetable matter on nematode populations. Plant Dis. Rep. 44:400–401.

Hutchinson, M. T., J. P. Reed, H. T. Streu, A. A. DiEdwardo, and P. H. Schroeder. 1961. Plant parasitic nematodes of New Jersey. New Jersey Agric. Exp. Sta. Bull. 796. 33 p.

Hutchinson, S. A., and W. F. Mai. 1954. A study of the efficiency of the catching organs of *Dactylaria eudermata* (Drechs.) in relation to *Heterodera rostochiensis* (Wr.) in soil. Plant Dis. Rep. 38:185–186.

Hutton, D. G., R. E. Wilkinson, and W. F. Mai. 1973. Effect of two plant-parasitic nematodes on Fusarium dry root rot of beans. Phytopathology 63:749–751.

Hwang, S-W. 1970. Freezing and storage of nematodes in liquid nitrogen. Nematologica 16:305–308.

Ichinohe, M. 1972. Nematode diseases of rice. Pp. 127–143 *in* J. M. Webster, Ed., Economic nematology. Academic, New York.

Inagaki, H., and N. T. Powell. 1969. Influence of the root-lesion nematode on black shank symptom development in flue-cured tobacco. Phytopathology 59:1350–1355.

Irvine, W. A. 1966. The development of *Meloidogyne hapla* in alfalfa roots under controlled and variable temperatures. Iowa State J. Sci. 41:55–75.

Ishibashi, N. 1965. The increase in male adults by gamma-ray irradiation in the root-knot nematode, *Meloidogyne incognita* Chitwood. Nematologica 11:361–369.

Jaccard, P. 1912. The distribution of the flora in the Alpine zone. New Phytologist 11:37–50.

Jackson, L. W. R. 1948. Deterioration of shortleaf pine roots caused by a parasitic nematode. Plant Dis. Rep. 32:192.

Jatala, P., H. J. Jensen, and S. A. Russell. 1974. *Pristionchus lheritieri* as a carrier of *Rhizobium japonicum*. J. Nematol. 6:130–131.

Jenkins, W. R. 1956. *Paratylenchus projectus*, new species (Nematoda, Criconematidae), with a key to the species of *Paratylenchus*. J. Wash. Acad. Sci. 46:296–298.

Jenkins, W. R., and B. W. Coursen, 1957. The effect of root-knot nematodes,

Meloidogyne incognita acrita and *M. hapla*, on Fusarium wilt of tomato. Plant Dis. Rep. 41:182–186.

Jenkins, W. R., D. P. Taylor, R. A. Rohde, and B. W. Coursen. 1957. Nematodes associated with crop plants in Maryland. Univ. Md. Agric. Exp. Sta. Bull. A-89. 25 p.

Jensen, H. J. 1967. Do saprozoic nematodes have a significant role in epidemiology of plant diseases? Plant Dis. Rep. 51:98–102.

Jensen, H. J., and S. R. Siemer. 1971. Protection of *Fusarium* and *Verticillium* propagules from selected biocides following ingestion by *Pristionchus lheritieri*. J. Nematol. 3:23–27.

Jensen, H. J., H. B. Howell, and W. D. Courtney. 1958. Grass seed nematode and production of bentgrass seed. Ore. Agric. Exp. Sta. Bull. 565. 8 p.

Johnson, A. W. 1970. Pathogenicity and interaction of three nematode species on six Bermudagrasses. J. Nematol. 2:36–41.

Johnson, A. W., and R. H. Littrell. 1970. Pathogenicity of *Pythium aphanidermatum* to chrysanthemum in combined inoculations with *Belonolaimus longicaudatus* or *Meloidogyne incognita*. J. Nematol. 2:255–259.

Johnson, A. W., and C. J. Nusbaum. 1968. The activity of *Tylenchorhynchus claytoni,* *Trichodorus christiei, Pratylenchus brachyurus, P. zeae,* and *Helicotylenchus dihystera* in single and multiple inoculations on corn and soybean. Nematologica 14:9. (Abstr.)

Johnson, A. W., and C. J. Nusbaum. 1970. Interactions between *Meloidogyne incognita, M. hapla,* and *Pratylenchus brachyurus* in tobacco. J. Nematol. 2:334–340.

Johnson, A. W., and W. M. Powell. 1968. Pathogenic capabilities of a ring nematode, *Criconemoides lobatum*, on various turf grasses. Plant Dis. Rep. 52:109–113.

Johnson, A. W., C. C. Dowler, and E. W. Hauser. 1974. Seasonal population dynamics of selected plant-parasitic nematodes on four monocultured crops. J. Nematol. 6:187–190.

Johnson, A. W., C. C. Dowler, and E. W. Hauser. 1975. Crop rotation and herbicide effects on population densities of plant-parasitic nematodes. J. Nematol. 7:158–168.

Johnson, E. C. 1909. Notes on a nematode in wheat. Science 30:576.

Johnson, L. F. 1959. Effect of the addition of organic amendments to soil on root knot of tomatoes. I. Preliminary report. Plant Dis. Rep. 43:1059–1062.

Johnson, L. F. 1962. Effect of addition of organic amendments to soil on root knot of tomatoes. II. Relation of soil temperature, moisture, and pH. Phytopathology 52:410–413.

Johnson, L. F. 1974. Extraction of oat straw, flax, and amended soil to detect substances toxic to the root-knot nematode. Phytopathology 64:1471–1473.

Johnson, L. F., and N. B. Shamiyeh. 1968. Evaluation of an agar-slide technique for studying egg hatching of *Meloidogyne incognita* in soil. Phytopathology 58:1665–1668.

Johnson, L. F., A. Y. Chambers, and H. E. Reed. 1967. Reduction of root knot of tomatoes with crop residue amendments in field experiments. Plant Dis. Rep. 51:219–222.

Johnson, R. N., and D. R. Viglierchio. 1961. The accumulation of plant parasitic nematode larvae around carbon dioxide and oxygen. Proc. Helminthol. Soc. Wash. 28:171–174.

Johnson, R. N., and D. R. Viglierchio. 1969. Sugar beet nematode (*Heterodera schachtii*) reared on axenic *Beta vulgaris* root explants. I. Selected environmental factors affecting penetration. Nematologica 15:129–143.

Johnson, S. R., V. R. Ferris, and J. M. Ferris. 1972. Nematode community structure of forest woodlots. I. Relationships based on similarity coefficients of nematode species. J. Nematol. 4:175–183.

Johnson, S. R., J. M. Ferris, and V. R. Ferris. 1973. Nematode community structure in forest woodlots. II. Ordination of nematode communities. J. Nematol. 5:95–107.

Johnson, S. R., J. M. Ferris, and V. R. Ferris. 1974. Nematode community structure of forest woodlots. III. Ordinations of taxonomic groups and biomass. J. Nematol. 6:118–126.

Johnston, T. M. 1957. Further studies on microbiological reduction of nematode population in water-saturated soils. Phytopathology 47:525–526. (Abstr.)

Johnston, T. M. 1959. Effect of fatty acid mixtures on the rice stylet nematode (*Tylenchorhynchus martini* Fielding, 1956). Nature 183:1392.

Jones, F. G. W. 1975. Accumulated temperature and rainfall as measures of nematode development and activity. Nematologica 21:62–70.

Jones, F. G. W., and A. J. Thomasson. 1976. Bulk density as an indicator of pore space in soils usable by nematodes. Nematologica 22:133–137.

Jones, F. G. W., D. W. Larbey, and D. M. Parrott. 1969. The influence of soil structure and moisture on nematodes, especially *Xiphinema, Longidorus, Trichodorus,* and *Heterodera* spp. Soil Biol. and Biochem. 1:153–165.

Jorgenson, E. C. 1970. Antagonistic interaction of *Heterodera schachtii* Schmidt and *Fusarium oxysporum* (Woll.) on sugarbeets. J. Nematol. 2:393–398.

Kable, P. F., and W. F. Mai. 1968a. Influence of soil moisture on *Pratylenchus penetrans*. Nematologica 14:101–122.

Kable, P. F., and W. F. Mai. 1968b. Overwintering of *Pratylenchus penetrans* in a sandy loam and a clay loam soil at Ithaca, New York. Nematologica 14:150–152.

Kalinenko, V. O. 1936. The inoculation of phytopathogenic microbes into rubber-bearing plants by nematodes. Phytopathol. Zeitsch. 9:407–416.

Katznelson, H., D. C. Gillespie, and F. D. Cook. 1964. Studies on the relationships between nematodes and other soil microorganisms. III. Lytic action of soil myxobacters on certain species of nematodes. Can. J. Microbiol. 10:699–704.

Kerstan, U. 1969. Die Beeinflussung des Geschlechterverhältnisses in der Gattung *Heterodera*. II. Minimallebensraum-selektive absterberate der geschlechter-geschlechterverhältnis (*Heterodera schachtii*). Nematologica 15:210–228.

Ketudat, U. 1969. The effects of some soil-borne fungi on the sex ratio of *Heterodera rostochiensis* on tomato. Nematologica 15:229–233.

Khan, M. H. 1972. Economic aspects of crop losses and disease control. Pp. 1–16 *in* J. M. Webster, Ed., Economic nematology. Academic, New York.

Khera, S., and B. M. Zuckerman. 1962. Studies on the culturing of certain ectoparasitic nematodes on plant callus tissue. Nematologica 8:272–274.

Khera, S., and B. M. Zuckerman. 1963. In vitro studies of host–parasite relationships of some plant-parasitic nematodes. Nematologica 9:1–6.

Khera, S., G. C. Bhatnagar, N. Kumar, and M. G. Tikyani. 1968. Studies on the culturing of Ditylenchus myceliophagus Goodey, 1958. Indian Phytopathol. 21:103–106.

Khoury, F. Y., and S. M. Alcorn. 1973. Effect of Meloidogyne incognita acrita on the susceptibility of cotton plants to Verticillium albo-atrum. Phytopathology 63:485–490.

Kimpinski, J., and H. E. Welch. 1971. The ecology of nematodes in Manitoba soils. Nematologica 17:308–318.

Kimpinski, J., H. R. Wallace, and R. B. Cunningham. 1976. Influence of some environmental factors on populations of Pratylenchus minyus in wheat. J. Nematol. 8:310–314.

Kincaid, R. R., and N. Gammon, Jr. 1957. Effect of soil pH on the incidence of three soil-borne diseases of tobacco. Plant Dis. Rep. 41:177–179.

Kincaid, R. R., F. G. Martin, N. Gammon, Jr., H. L. Breland, and W. L. Pritchett. 1970. Multiple regression of tobacco black shank, root knot, and coarse root indexes on soil pH, potassium, calcium, and magnesium. Phytopathology 60:1513–1516.

Kinloch, R. A., and M. W. Allen. 1972. Interaction of Meloidogyne hapla and M. javanica infecting tomato. J. Nematol. 4:7–16.

Kirkpatrick, J. D., W. F. Mai, K. G. Parker, and E. G. Fisher. 1964. Effect of phosphorus and potassium nutrition on sour cherry on the soil population levels of five plant-parasitic nematodes. Phytopathology 54:706–712.

Kiesiel, M., K. Deubert, and B. M. Zuckerman. 1969. The effect of Tylenchus agricola and Tylenchorhynchus claytoni on root rot of corn caused by Fusarium roseum and Pythium ultimum. Phytopathology 59:1387–1390.

Klekowski, R. Z., L. Wasilewska, and E. Paplinska. 1972. Oxygen consumption by soil-inhibiting nematodes. Nematologica 18:391–403.

Klingler, J. 1961. Anziehungsversuch mit Ditylenchus dipsaci unter Berücksichtigung der Wirkung des Kohlendioxyds, des Redoxpotentials und anderer Faktoren. Nematologica 6:69–84.

Klingler, J. 1963. Die orientierung von Ditylenchus dipsaci in Gemessenen Kunstlichen und Biologischen CO_2-gradienten. Nematologica 9:185–199.

Klingler, J. 1965. On the orientation of plant nematodes and of some other soil animals. Nematologica 11:4–18.

Klink, J. W., V. H. Dropkin, and J. E. Mitchell. 1970. Studies on the host-finding mechanisms of Neotylenchus linfordi. J. Nematol. 2:106–117.

Klinkenberg, C. H. 1963. Observations on the feeding habits of Rotylenchus uniformis, Pratylenchus crenatus, P. penetrans, Tylenchorhynchus dubius, and Hemicycliophora similis. Nematologica 9:502–506.

Knobloch, N. A. 1975. Quinisulcius tarjani sp. n. (Nematoda: Tylenchorhynchinae) with key to Quinisulcius species and notes on other plant-parasitic nematodes from Mexico. Proc. Helminthol. Soc. Wash. 42:52–56.

Knobloch, N. A., and C. W. Laughlin. 1973. A collection of plant parasitic nematodes (Nematoda) from Mexico with descriptions of three new species. Nematologica 19:205–217.

Krebill, R. G., K. R. Barker, and R. F. Patten. 1967. Plant-parasitic nematodes of Jack and red pine stands in Wisconsin. Nematolgica 13:33–42.

Krusberg, L. R. 1959. Investigations on the life cycle, reproduction, feeding habits and host range of *Tylenchorhynchus claytoni* Steiner. Nematologica 4:187–197.

Krusberg, L. R., and J. N. Sasser. 1956. Host-parasite relationships of the lance nematode in cotton roots. Phytopathology 46:505–510.

Lamberti, F. 1969. Effect of temperature on the reproduction rate of *Longidorus africanus*. Plant Dis. Rep. 53:559.

Laughlin, C. W., and A. S. Williams. 1971. Population behavior of *Meloidogyne graminis* in field-grown 'Tifgreen' Bermudagrass. J. Nematol. 3:386–389.

Laughlin, C. W., A. S. Williams, and J. A. Fox. 1969. The influence of temperature on development and sex differentiation of *Meloidogyne graminis*. J. Nematol. 1:212–215.

Leake, P. A., and H. J. Jensen. 1970. Survival of Chlorophyceae ingested by saprozoic nematodes. J. Nematol. 2:351–354.

Lear, B. 1959. Application of castor pomace and cropping of castor beans to soil to reduce nematode populations. Plant Dis. Rep. 43:459–460.

Lear, B., D. E. Johnson, and S. T. Miyagawa. 1969. A disease of lettuce associated with an ectoparasitic nematode, *Rotylenchus robustus*. Plant Dis. Rep. 53:952–954.

Lee, D. L. 1965. The physiology of nematodes. Oliver and Boyd, Edinburgh.

Leukel, R. W. 1924. Investigations on the nematode disease of cereals caused by *Tylenchus tritici*. J. Agric. Res. 27:925–956.

Lewis, F. J., and W. F. Mai. 1960. Survival of encysted and free larvae of the golden nematode in relation to temperature and humidity. Proc. Helminthol. Soc. Wash. 27:80–85.

Lewis, G. D., and W. F. Mai. 1960. Overwintering and migration of *Ditylenchus dipsaci* in organic soils of southern New York. Phytopathology 50:341–343.

Limber, D. P. 1973. Notes on the longevity of *Anguina tritici* (Steinbuch, 1799) Filipjev, 1936, and its ability to invade wheat seedlings after thirty-two years of dormancy. Proc. Helminthol. Soc. Wash. 40:272–274.

Linford, M. B. 1937. The feeding of some hollow-stylet nematodes. Proc. Helminthol. Soc. Wash. 4:41–46.

Linford, M. B. 1939. Attractiveness of roots and excised shoot tissue to certain nematodes. Proc. Helminthol. Soc. Wash. 6:11–18.

Linford, M. B., and J. M. Oliveira. 1937. The feeding of hollow-spear nematodes on other nematodes. Science 85:295–297.

Linford, M. B., and J. M. Oliveira. 1938. Potential agents of biological control of plant-parasitic nematodes. Phytopathology 28:14. (Abstr.)

Linford, M. B., and J. M. Oliveira. 1940. *Rotylenchulus reniformis*, nov. gen., n. sp., a nematode parasite of roots. Proc. Helminthol. Soc. Wash. 7:35–42.

Linford, M. B., and F. Yap. 1939. Root-knot nematode injury restricted by a fungus. Phytopathology 29:596–609.

Linford, M. B., F. Yap, and J. M. Oliveira. 1938. Reductions of soil populations of root-knot nematode during decomposition of organic matter. Soil Sci. 45:127–141.

Linford, M. B., J. M. Oliveira, and M. Ishi. 1949. *Paratylenchus minutus,* n. sp., a nematode parasitic on roots. Pacific Sci. 3:111–119.

Loewenberg, J. R., T. Sullivan, and M. L. Schuster. 1959. A virus disease of *Meloidogyne incognita incognita,* the southern root knot nematode. Nature 184:1896.

Loof, P. A. A. 1960. Taxonomic studies on the genus *Pratylenchus* (Nematoda). Tijdschr. Plantenz. 66:29–90.

Loof, P. A. A. 1971. Free-living and plant parasitic nematodes from Spitzbergen, collected by Mr. H. van Rossen. Mededel. Landlow. 71–7. 86 p.

Loos, C. A. 1959. Symptom expression of Fusarium wilt disease of the Gros Michel banana in the presence of *Radopholus similis* (Cobb, 1893) Thorne, 1949 and *Meloidogyne incognita acrita* Chitwood, 1949. Proc. Helminthol. Soc. Wash. 26:103–111.

Loos, C. A. 1962. Studies on the life-history and habits of the burrowing nematode, *Radopholus similis,* the cause of black-head disease of banana. Proc. Helminthol. Soc. Wash. 29:43–52.

Lownsbery, B. F. 1959. Studies of the nematode, *Criconemoides xenoplax,* on peach. Plant Dis. Rep. 43:913–917.

Lownsbery, B. F. 1961. Factors affecting population levels of *Criconemoides xenoplax.* Phytopathology 51:101–103.

Lownsbery, B. F., and A. R. Maggenti. 1963. Some effects of soil temperature and soil moisture on population levels of *Xiphinema americanum.* Phytopathology 53:667–668.

Lownsbery, B. F., and D. R. Viglierchio. 1960. Mechanism of accumulation of *Meloidogyne incognita acrita* around tomato seedlings. Phytopathology 50:178–179.

Lownsbery, B. F., H. English, E. H. Moody, and F. J. Shick. 1973. *Criconemoides xenoplax* experimentally associated with a disease of peach. Phytopathology 63:994–997.

Luc, M. 1961. Note préliminaire sur le déplacement de *Hemicycliophora paradoxa* Luc (Nematoda–Criconematidae) dans le sol. Nematologica 6:95–106.

Machmer, J. H. 1958. Effect of soil salinity on nematodes in citrus and papaya plantings. J. Rio Grande Valley Hort. Soc. 12:57–60.

Maggenti, A. R. 1971. Nemic relationships and the origins of plant parasitic nematodes. Pp. 65–81 *in* B. M. Zuckerman, W. F. Mai, and R. A. Rohde, Eds., Plant parasitic nematodes. Vol. 1. Academic, New York.

Maggenti, A. R., and A. Hardan. 1973. The effects of soil salinity and *Meloidogyne javanica* on tomato. J. Nematol. 5:231–234.

Maggenti, A. R., W. H. Hart, and G. A. Paxman. 1973. A new genus and species of gall forming nematode from *Danthonia californica,* with discussion of its life history. Nematologica 19:491–497.

Mai, W. F., and K. G. Parker. 1956. Evidence that the nematode *Pratylenchus penetrans* causes losses in New York State cherry orchards. Phytopathology 46:19. (Abstr.)

Mai, W. F., H. W. Crittenden, and W. R. Jenkins. 1960. Distribution of stylet-bearing nematodes in the northeastern United States. New Jersey Agric. Exp. Sta. Bull. 795. 62 p.

Malek, R. B. 1969. Population fluctuations and observations of the life cycle of *Xiphinema americanum* associated with cottonwood (*Populus deltoides*) in South Dakota. Proc. Helminthol. Soc. Wash. 36:270–274.

Malek, R. B., and J. B. Gartner. 1975. Hardwood bark as a soil amendment for suppression of plant parasitic nematodes on container-grown plants. Hort Science 10:33–35.

Malek, R. B., and W. R. Jenkins. 1964. Aspects of the host–parasitic relationships of nematodes and hairy vetch. New Jersey Agric. Exp. Sta. Bull 813. 31 p.

Mamiya, Y. 1971. Effect of temperature on the life cycle of *Pratylenchus penetrans* on *Cryptomera* seedlings and observations on its reproduction. Nematologica 17:82–92.

Mamiya, Y., and N. Enda. 1972. Transmission of *Bursaphelenchus lignicolus* (Nematoda: Aphelenchoididae) by *Monochamus alternatus* (Coleoptera: Cerambycidae). Nematologica 18:159–162.

Mamiya, Y., and T. Kiyohara. 1972. Description of *Bursaphelenchus lignicolus*, n. sp. (Nematoda: Aphelenchoididae) from pine wood and histopathology of nematode-infested trees. Nematologica 18:120–124.

Mankau, R. 1961a. An attempt to control root-knot nematode with *Dactylaria thaumasia* and *Arthrobotrys arthrobotryoides*. Plant Dis. Rep. 45:164–166.

Mankau, R. 1961b. The use of nematode-trapping fungi to control root-knot nematodes. Nematologica 6:326–332.

Mankau, R. 1962. Soil fungistasis and nematophagus fungi. Phytopathology 52:611–615.

Mankau, R. 1968. Reduction of root-knot disease with organic amendments under semifield conditions. Plant Dis. Rep. 52:315–319.

Mankau, R. 1969. Nematicidal activity of *Aspergillus niger* culture filtrates. Phytopathology 59:1170.

Mankau, R. 1975. *Bacillus penetrans* n. comb. causing a virulent disease of plant-parasitic nematodes. J. Invert. Pathol. 26:333–339.

Mankau, R., and S. Das. 1969. The influence of chitin amendments on *Meloidogyne incognita*. J. Nematol. 1:15–16. (Abstr.)

Mankau, R., and M. B. Linford. 1960. Host–parasite relationships of the clover cyst nematode, *Heterodera trifolii* Goffart. Univ. Ill. Agric. Exp. Sta. Bull. 667. 50 p.

Mankau, R., and R. J. Minteer. 1962. Reduction of soil populations of the citrus nematode by the addition of organic materials. Plant Dis. Rep. 46:375–378.

Mankau, R., and N. Prasad. 1977. Infectivity of *Bacillus penetrans* in plant-parasitic nematodes. J. Nematol. 9:40–45.

Marchant, E. H. J. 1934. The estimated number of nemas in the soils of Manitoba. Can. J. Res. 11:594–601.

Marcinowski, K. 1910. Parasitisch und semiparasitisch an Pflanzen lebende Nematoden. Berlin Kaiserlichen Biologischen Anstalt fur Land und Forstwirtschaft. 7:1–192.

Marks, C. F., and R. M. Sayre. 1964. The effect of potassium on the rate of development of the root-knot nematodes *Meloidogyne incognita, M. javanica* and *M. hapla.* Nematologica 10:323–327.

Marlatt, R. B. 1970. Transmission of *Aphelenchoides besseyi* to *Ficus elastica* leaves via *Sporobolus poiretii* inflorescences. Phytopathology 60:543–544.

Martin, G. C. 1968a. Control of *Meloidogyne javanica* in potato tubers and *M. hapla* in the roots of young rose bushes by means of heated water. Nematologica 14:441–446.

Martin, G. C. 1968b. Survival and infectivity of eggs and larvae of *Meloidogyne arenaria* after ingestion by a bovine. Plant Dis. Rep. 52:99–102.

Martin, G. C. 1969. Survial and infectivity of eggs and larvae of *Meloidogyne javanica* after ingestions by a rodent. Nematologica 15:620.

Martin, W. J., L. D. Newsom, and J. E. Jones. 1956. Relationship of nematodes to the development of Fusarium wilt in cotton. Phytopathology 46:285–289.

Mayol, P. S., and G. B. Bergeson. 1970. The role of secondary invaders in *Meloidogyne incognita* infection. J. Nematol. 2:80–83.

McClellan, W. D., and J. R. Christie. 1949. Incidence of *Fusarium* infection as affected by root-knot nematodes. Phytopathology 39:568–571;

McClure, M. A., and D. R. Viglierchio. 1966. The influence of host nutrition and intensity of infection on the sex ratio and development of *Meloidogyne incognita* in sterile agar cultures of excised cucumber roots. Nematologica 12:248–258.

McElroy, F. D. 1970. *Xiphinema bakeri,* a nematode pest of raspberry. J. Parasitol. 56:448. (Abstr.)

McElroy, F. D., and S. D. Van Gundy. 1968. Observations on the feeding processes of *Hemicycliophora arenaria.* Phytopathology 58:1558–1565.

McElroy, F. D., S. A. Sher, and S. D. Van Gundy. 1966. The sheath nematode, *Hemicycliophora arenaria,* a native to California soils. Plant Dis. Rep. 50:581–583.

McGlohon, N. E., J. N. Sasser, and R. T. Sherwood. 1961. Investigations of plant-parasitic nematodes associated with forage crops in North Carolina. N. Car. Agric. Exp. Sta. Tech. Bull. 148. 39 p.

McGuire, J. M. 1973. Retention of tobacco ringspot virus by *Xiphinema americanum.* Phytopathology 63:324–326.

McLeod, R. W. 1968. The effects of *Aphelenchoides saprophilus, A. coffeae, Aphelenchus avenae,* and *Panagrolaimus* sp. on the cropping of cultivated mushroom. Nematologica 14:573–576.

Meagher, J. W. 1968. The distribution of the cereal cyst nematode (*Heterodera avenae*) in Victoria and its relation to soil type. Austral. J. Exp. Agric. and Animal Husb. 8:637–640.

Meagher, J. W. 1974. Cryptobiosis of the cereal cyst nematode (*Heterodera avenae*), and effects of temperature and relative humidity on survival of eggs in storage. Nematologica 20:323–335.

Meagher, J. W., and S. C. Chambers. 1971. Pathogenic effects of *Heterodera avenae* and *Rhizoctonia solani* and their interaction on wheat. Austral. J. Agric. Res. 22:189–194.

Meagher, J. W., and P. T. Jenkins. 1970. Interaction of *Meloidogyne hapla* and *Verticillium dahliae*, and the chemical control of wilt in strawberry. Austral. J. Exp. Agric. and Animal Husb. 10:493–496.

Meagher, J. W., and D. R. Rooney, 1966. The effect of crop rotations in the Victorian Wimmera on the cereal cyst nematode (*Heterodera avenae*), nitrogen fertility and wheat yield. Austral. J. Exp. Agric. and Animal Husb. 6:425–431.

Michell, R. E., and W. M. Powell. 1972. Influence of *Pratylenchus brachyurus* on the incidence of Fusarium wilt in cotton. Phytopathology 62:336–338.

Micoletzky, H. 1922. Die freilebenden Nematoden. Arch. Naturgesch. (1921) 87.A:1–650.

Micoletzky, H. 1925. Die freilebenden Süsswasser- und Moornematoden Dänemarks Nebst Anhang: Uber Amöbosporidien und andere Parasiten bei freilebenden Nemtoden. K. danske Vidensk. Selek Skr. 8th Ser. 10(2):57–310.

Miller, L. I. 1972. The influence of soil texture on the survival of *Belonolaimus longicaudatus*. Phytopathology 62:670–671. (Abstr.)

Miller, P. M. 1968. The susceptibility of parasitic nematodes to sub-freezing temperatures. Plant Dis. Rep. 52:768–772.

Miller, P. M. 1969. Injury to *Heterodera tabacum* by high soil temperatures. Plant Dis. Rep. 53:191–193.

Miller, P. M. 1970. Rate of increase of a low population of *Heterodera tabacum* reduced by *Pratylenchus prentrans* in the soil. Plant Dis. Rep. 54:25–26.

Miller, P. M. 1975. Effect of the tobacco cyst nematode, *Heterodera tabacum*, on severity of Verticillium and Fusarium wilts of tomato. Phytopathology 65:81–82.

Miller, P. M., and S. E. Wihrheim. 1968. Mutual antagonism between *Heterodera tabacum* and some other parasitic nematodes. Plant Dis. Rep. 52:57–58.

Miller, P. M., G. S. Taylor, and S. E. Wihrheim. 1968. Effects of cellulosic soil amendments and fertilizers on *Heterodera tabacum*. Plant Dis. Rep. 52:441–445.

Miller, P. M., D. C. Sands, and S. Rich. 1973. Effect of industrial mycelial residues, wood fiber wastes, and chitin on plant-parasitic nematodes and some soilborne diseases. Plant Dis. Rep. 57:438–442.

Miller, R. E., C. W. Boothroyd, and W. F. Mai. 1963. Relationship of *Pratylenchus penetrans* to roots of corn in New York. Phytopathology 53:313–315.

Milne, D. L., and D. P. DuPlessis. 1964. Development of *Meloidogyne javanica* (Treub.) Chit., on tobacco under fluctuating soil temperatures. S. Afr. J. Agric. Sci. 7:673–680.

Minton, N. A., and D. K. Bell. 1969. *Criconemoides ornatus* parasitic on peanuts. J. Nematol. 1:349–351.

Minton, N. A., and M. B. Parker. 1975. Interaction of four soybean cultivars with subsoiling and a nematicide. J. Nematol. 7:60–64.

Minton, N. A., E. J. Cairns, and A. L. Smith. 1960. Effect on root-knot nematode populations of resistant and susceptible cotton. Phytopathology 50:784–787.

Minton, N. A., D. K. Bell, and B. Doupnik, Jr. 1969. Peanut pod invasion by *Aspergillus flavus* in the presence of *Meloidogyne hapla*. J. Nematol. 1:318–320.

Miura, K. 1966. Nutrition and spontaneous trap-formation of *Arthrobotrys dactyloides* Drechsler. Trans. Mycol. Soc. Japan 7:320–324.

Miyagawa, S. T., and B. Lear. 1970. Factors influencing survival of *Ditylenchus dipsaci* (Kühn, 1857) in soil. J. Nematol. 2:139–146.

Mojtahedi, H., B. F. Lownsbery, and E. H. Moody. 1975. Ring nematodes increase development of bacterial cankers in plums. Phytopathology 65:556–559.

Monoson, H. L. 1968. Trapping effectiveness of five species of nematophagous fungi cultured with mycophagous nematodes. Mycologia 60:788–801.

Monoson, H. L., A. G. Galsky, and R. S. Stephano. 1974. Studies on the ability of various nematodes to induce trap formation in a nematode-trapping fungus *Monocrosporium doedycoides*. Nematologica 20:96–102.

Morgan, G. T., and A. A. MacLean. 1968. Influence of soil pH on an introduced population of *Pratylenchus penetrans*. Nematologica 14:311–312.

Morris, R. F. 1971. Distribution and biology of the golden nematode *Heterodera rostochiensis* in Newfoundland. Nematologica 17:370–376.

Mountain, W. B. 1954. Studies of nematodes in relation to brown root rot of tobacco in Ontario. Can. J. Bot. 32:737–759.

Mountain, W. B., and H. R. Boyce. 1958. The peach replant problem in Ontario. V. The relation of parasitic nematodes to regional differences in severity of peach replant failure. Can. J. Bot. 36:125–134.

Mountain, W. B., and C. D. McKeen. 1962. Effect of *Verticillium dahliae* on the population of *Pratylenchus penetrans*. Nematologica 7:261–266.

Mountain, W. B., and Z. A. Patrick. 1959. The peach replant problem in Ontario. VII. The pathogenicity of *Pratylenchus penetrans* (Cobb, 1917) Filip. & Stek. 1941. Can. J. Bot. 37:459–470.

Moura, de, R. M., E. Echandi, and N. T. Powell. 1975. Interaction of *Corynebacterium michiganense* and *Meloidogyne incognita* on tomato. Phytopathology 65:1332–1335.

Moussa, M. T. 1969. Nematode fossil tracks of Eocene age from Utah. Nematologica 15:376–380.

Mulvey, R. H. 1963. A grass nematode, *Anguina agrostis* (Steinbuch), on *Arctagrostis latifolia* from the Canadian arctic. Can. J. Zool. 41:1223–1226.

Mulvey, R. H. 1969a. Nematodes of the family Neotylenchidae (Tylenchida: Nematoda) from the Canadian high arctic. Can. J. Zool. 47:1261–1268.

Mulvey, R. H. 1969b. Nematodes of the genus *Tylenchorhynchus* (Tylenchoidea: Nematoda) from the Canadian high arctic. Can. J. Zool. 47:1245–1248.

Murad, J. L. 1970. Population study of nematodes from drying beds. Proc. Helminthol. Soc. Wash. 37:10–13.

Murphy, P. W., and C. C. Doncaster. 1957. A culture method for soil meiofauna and its application to the study of nematode predators. Nematologica 2:202–214.

Myers, R. F. 1965. Amylase, cellulase, invertase and pectinase in several free-living, mycophagus, and plant-parasitic nematodes. Nematologica 11:441–448.

Myers, R. F., and L. R. Krusberg. 1965. Organic substances discharged by plant-parasitic nematodes. Phytopathology 55:429–437.

National Academy of Science. 1968. Principles of plant and animal pest control. Vol. 4. Control of plant-parasitic nematodes. Pub. 1696. Nat. Acad. Sci. 172 p.

Neal, D. C. 1954. The reniform nematode and its relationship to the incidence of Fusarium wilt of cotton at Baton Rouge, Louisiana. Phytopathology 44:447–450.

Nickle, W. R. 1960. Nematodes associated with the rootlets of western white pine in northern Idaho. Plant Dis. Rep. 44:470–471.

Nikandrow, A., and C. D. Blake. 1972. Oxygen and the hatch of eggs and development of larvae of *Aphelenchoides composticola* and *Ditylenchus myceliophagus*. Nematologica 18:309–319.

Nishizawa, T., and K. Iwatomi. 1955. *Nothotylenchus acris* Thorne as a parasitic nematode of strawberry plant. Jap. J. Appl. Zool. 20:47–55.

Norgren, R. L., B. M. Zuckerman, and D. H. MacDonald. 1968. Investigations on the biology and monoxenic culturing of *Tetylenchus joctus*. Nematologica 14:565–572.

Norton, D. C. 1959. Relationship of nematodes to small grains and native grasses in north and central Texas. Plant Dis. Rep. 43:227–235.

Norton, D. C. 1963. Population fluctuations of *Xiphinema americanum* in Iowa. Phytopathology 53:66–68.

Norton, D. C. 1965a. *Anguina agropyronifloris* n. sp., infecting florets of *Agropyron smithii*. Proc. Helminthol. Soc. Wash. 32:118–122.

Norton, D. C. 1965b. *Xiphinema americanum* populations and alfalfa yields as affected by soil treatment, spraying, and cutting. Phytopathology 55:615–619.

Norton, D. C. 1967. Relationships of *Heterodera trifolii* to some forage legumes. Phytopathology 57:1305–1308.

Norton, D. C. 1969. *Meloidogyne hapla* as a factor in alfalfa decline in Iowa. Phytopathology 59:1824–1828.

Norton, D. C., and N. Burns. 1971. Colonization and sex ratios of *Pratylenchus alleni* in soybean roots under two soil moisture regimes. J. Nematol. 3:374–377.

Norton, D. C., and J. K. Hoffmann. 1974. Distribution of selected plant parasitic nematodes relative to vegetation and edaphic factors. J. Nematol. 6:81–86.

Norton, D. C., and J. K. Hoffmann. 1975. *Longidorus breviannulatus* n. sp. (Nematoda: Longidoridae) associated with stunted corn in Iowa. J. Nematol. 7:168–171.

Norton, D. C., and J. E. Sass. 1966. Pathological changes in *Agropyron smithii* induced by *Anguina agropyronifloris*. Phytopathology 56:769–771.

Norton, D. C., W. A. Irvine, and D. Castaner. 1964. Nematodes associated with plants in Iowa. Proc. Iowa Acad. Sci. 71:109–116.

Norton, D. C., L. R. Frederick, P. E. Ponchillia, and J. W. Nyhan. 1971. Correlations of nematodes and soil and properties in soybean fields. J. Nematol. 3:154–163.

Nusbaum, C. J., and H. Ferris. 1973. The role of cropping systems in nematode population management. Ann. Rev. Phytopathol. 11:423–440.

O'Bannon, J. H. 1968. The influence of an organic soil amendment on infectivity and reproduction of *Tylenchulus semipenetrans* on two citrus rootstocks. Phytopatholoy 58:597–601.

O'Bannon, J. H., and H. W. Reynolds. 1961. Root-knot nematode damage and cotton yields in relation to certain soil properties. Soil Sci. 92:384–386.

O'Bannon, J. H., C. R. Leathers, and H. W. Reynolds. 1967. Interactions of *Tylenchuls semipenetrans* and *Fusarium* species on rough lemon (*Citrus limon*). Phytopathology 57:414–417.

O'Bannon, J. H., J. D. Radewald, and A. T. Tomerlin. 1972. Population fluctuation of three parasitic nematodes in Florida citrus. J. Nematol. 4:194–199.

Olthof, T. H. A. 1971. Seasonal fluctuations in population densities of *Pratylenchus penetrans* under a rye-tobacco rotation in Ontario. Nematologica 17:453–459.

Olthof, T. H. A., and R. H. Estey. 1963. A nematotoxin produced by the nematophagous fungus *Arthrobotrys oligospora* Fresenius. Nature 197:514–515.

Olthof, T. H. A., and R. H. Estey. 1965. Relation of some environmental factors to growth of several nematophagous Hyphomycetes. Can. J. Microbiol. 11:939–946.

Olthof, T. H. A., and R. H. Estey. 1966. Carbon and nitrogen levels of a medium in relation to growth and nematophagous activity of *Arthrobotrys oligospora* Fresenius. Nature 209:1158.

Oostenbrink, M. 1966. Major characteristics of the relation between nematodes and plants. Meded. Landbouwhogeschool Wageningen 66-4. 46 p.

Oostenbrink, M. 1971. Quantitative aspects of plant–nematode relationships. Indian J. Nematol. 1:68–74.

Oostenbrink, M. J. J. s'Jacob, and K. Kuiper. 1956. An interpetation of some crop rotation experiences based on nematode surveys and population studies. Nematologica 1:202–215.

Orion, D., and G. Minz. 1971. The influence of morphactin on the root-knot nematode *Meloidogyne javanica* and its galls. Nematologica 17:107–112.

Orr, C. C., and O. J. Dickerson. 1966. Nematodes in true prairie soils of Kansas. Trans. Kans. Acad. Sci. 69:317–334.

Orr, C. C., and O. H. Newton. 1971. Distribution of nematodes by wind. Plant Dis. Rep. 55:61–63.

Oteifa, B. A. 1951. Effects of potassium nutrition and amount of inoculum on rate of reproduction of *Meloidogyne incognita*. J. Wash. Acad. Sci. 41:393–395.

Oteifa, B. A. 1952. Potassium nutrition of the host in relation to infection by a root-knot nematode *Meloidogyne incognita*. Proc. Helminthol. Soc. Wash. 19:99–104.

Oteifa, B. A. 1953. Development of the root-knot nematode, *Meloidogyne incognita*, as affected by potassium nutrition of the host. Phytopathology 43:171–174.

Oteifa, B. A. 1955. Nitrogen source of the host nutrition in relation to infection by a root-knot nematode, *Meloidogyne incognita*. Plant Dis. Rep. 39:902–903.

Oteifa, B. A., and M. F. Abdelhalim. 1957. Effects of soil nature and seasonal changes on nematode population of Geja (Egypt) soil. Cairo Fac. Agr. Bull. 129. 9 p.

Oteifa, B. A., and D. M. Elgindi. 1961. Physiological studies on host–parasite relationship of the root-knot nematode, *Meloidogyne javanica*. Plant Dis. Rep. 45:928–929.

Oteifa, B. A., D. M. Elgindi, and K. A. Diab. 1964. Cotton yield and population dynamics of the stunt nematode *Tylenchorhynchus latus* under mineral fertilizer trials. Potash Rev. 23:31.

Overgaard-Nielsen, C. 1948. Studies on the soil microfauna. I. The moss inhibiting nematodes and rotifers. Publ. Soc. Sci. Lettr. d'Aarbus. Ser. Sci. Nat. 1.

Overgaard-Nielsen, C. 1949. Studies on the soil microfauna. II. The soil inhabiting nematodes. Natura Jutland. 2:1–131.

Oyekan, P. O., and J. E. Mitchell. 1971. Effect of *Pratylenchus penetrans* on the resistance of a pea variety to Fusarium wilt. Plant Dis. Rep. 55:1032–1035.

Oyekan, P. O., and J. E. Mitchell. 1972. The role of *Pratylenchus penetrans* in the root rot complex of canning pea. Phytopathology 62:369–373.

Paesler, F. 1957. Beschreibung einiger nematoden aus champignonbeeten. Nematologica 2:314–328.

Palmer, A. R. 1974. Search for the Cambrian world. Amer. Sci. 62:216–224.

Palmer, L. T., and D. H. MacDonald. 1974. Interaction of *Fusarium* spp. and certain plant parasitic nematodes on maize. Phytopathology 64:14–17.

Paramonov. A. A. 1962. Plant-parasitic nematodes. I. Origin of nematodes, ecological and morphological characteristics of plant nematodes, principles of taxonomy. Translated from Russian by Israel Prog. Sci. Translations, 1968. U.S. Dept. Commerce. Springfield, Va.

Parker, K. G., and W. F. Mai. 1956. Damage to tree fruits in New York by root lesion nematodes. Plant Dis. Rep. 40:694–699.

Patel, K. P., and D. H. MacDonald. 1968. Winter survival of *Tylenchorhynchus martini* in Minnesota. Phytopathology 58:266–267.

Patrick, Z. A., R. M. Sayre, and H. J. Thorpe. 1965. Nematocidal substances selective for plant-parasitic nematodes in extracts of decomposing rye. Phytopathology 55:702–704.

Patterson, Sis. M. T., and G. B. Bergeson. 1967. Influence of temperature, photoperiod, and nutrition on reproduction, male-female-juvenile ratio, and root to soil migration of *Pratylenchus penetrans*. Plant Dis. Rep. 51:78–82.

Peacock, F. C. 1957. Studies on root knot nematodes of the genus *Meloidogyne* in the Gold Coast. Part II. The effect of soil moisture content on survival of the organism. Nematologica 2:114–122.

Pepper, E. H. 1963. Nematodes in North Dakota. Plant Dis. Rep. 47:102–106.

Perry, V. G. 1953. The awl nematode, *Dolichodorus heterocephalus*, a devastating plant parasite. Proc. Helminthol. Soc. Wash. 20:21–27.

Perry, V. G., H. M. Darling, and G. Thorne. 1959. Anatomy, taxonomy and control of certain spiral nematodes attacking blue grass in Wisconsin. Univ. Wis. Res. Bull. 207. 24 p.

Peters, B. G. 1926. *Heterodera schachtii* (Schmidt) and soil acidity. J. Helminthol. 4:87–114.

Petherbridge, F. R., and F. G. W. Jones. 1944. Beet eelworm (*Heterodera schachtii* Schm.) in East Anglia, 1934–1943. Ann. Appl. Biol. 31:320–332.

Pielou, E. C. 1972. Measurement of structure in animal communities. Pp. 113–135 *in* J. A. Wiens, Ed., Ecosystem structure and function. Oregon State University Press, Corvallis.

Pielou, E. C. 1974. Population and community ecology. Principles and methods. Gordon and Breach, New York.

Pielou, E. C. 1975. Ecological diversity. Wiley-Interscience, New York.

Pielou, E. C. 1977. Mathematical ecology. Wiley-Interscience, New York.

Pillai, J. K., and D. P. Taylor. 1967. Influence of fungi on host preference, host suitability, and morphometrics of five mycophagous nematodes. Nematologica 13:529–540.

Pitcher, R. S. 1963. Role of plant-parasitic nematodes in bacterial diseases. Phytopathology 53:35–39.

Pitcher, R. S., and J. E. Crosse. 1958. Studies in the relationship of eelworms and bacteria to certain plant diseases. II. Further analysis of the strawberry cauliflower disease complex. Nematologica 3:244–256.

Pitcher, R. S., and D. G. McNamara. 1970. The effect of nutrition and season of year on the reproduction of *Trichodorus viruliferus*. Nematologica 16:99–106.

Pitcher, R. S., Z. A. Patrick, and W. B. Mountain. 1960. Studies on the host–parasite relations of *Pratylenchus penetrans* (Cobb) to apple seedlings. I. Pathogenicity under sterile conditions. Nematologica 5:309–314.

Ponchillia, P. E. 1972. *Xiphinema americanum* as affected by soil organic matter and porosity. J. Nematol. 4:189–193.

Porter, D. M., and N. T. Powell. 1967. Influence of certain *Meloidogyne* species on Fusarium wilt development in flue-cured tobacco. Phytopathology 57:282–285.

Powell, N. T. 1971. Interactions between nematodes and fungi in disease complexes. Ann. Rev. Phytopathol. 9:253–274.

Powell, N. T., and C. K. Batten. 1969. Complexes in tobacco involving *Meloidogyne incognita, Fusarium oxysporum* f. sp. *nicotianae,* and *Alternaria tenuis*. Phytopathology 59:1044. (Abstr.)

Powell, N. T., and C. J. Nusbaum. 1960. The black shank–root-knot complex in flue-cured tobacco. Phytopathology 50:899–906.

Powell, N. T., P. L. Meléndez, and C. K. Batten. 1971. Disease complexes in tobacco involving *Meloidogyne incognita* and certain soil-borne fungi. Phytopathology 61:1332–1337.

Pramer, D., and N. R. Stoll. 1959. Nemin: A morphogenic substance causing trap formation by predaceous fungi. Science 129:966–967.

Prasad, N., and R. Mankau. 1969. Studies on a sporozoan endoparasite of nematodes. J. Nematol. 1:301–302. (Abstr.)

Proctor, J. R., and C. F. Marks. 1974. The determination of normalizing transformations for nematode count data from soil samples and of efficient sampling schemes. Nematologica 20:395–406.

Putnam, D. F., and L. J. Chapman. 1935. Oat seedling diseases in Ontario. I. The oat nematode *Heterodera schachtii* Schm. Sci. Agric. 15:633–651.

Radewald, J. D., and D. J. Raski. 1962a. A study of the life cycle of *Xiphinema index*. Phytopathology 52:748. (Abstr.)

Radewald, J. D., and D. J. Raski. 1962b. Studies on the host range and pathogenicity of *Xiphinema index*. Phytopathology 52:748–749. (Abstr.)

Radewald, J. D., J. W. Osgood, K. S. Mayberry, A. O. Paulus, and F. Shibuya. 1969. *Longidorus africanus* a pathogen of head lettuce in the Imperial Valley of southern California. Plant Dis. Rep. 53:381–384.

Radewald, J. D., L. Pyeatt, F. Shibuya, and W. Humphrey. 1970. *Meloidogyne naasi*, a parasite of turfgrass in southern California. Plant Dis. Rep. 54:940–942.

Radewald, J. D., J. H. O'Bannon, and A. T. Tomerlin. 1971. Temperature effects on reproduction and pathogenicity of *Pratylenchus coffeae* and *P. brachyurus* and survival of *P. Coffeae* in roots of *Citrus jambhiri*. J. Nematol. 3:390–394.

Raski, D. J. 1952. On the morphology of *Criconemoides* Taylor, 1936, with descriptions of six new species (Nematoda:Criconematidae). Proc. Helminthol. Soc. Wash. 19:85–99.

Raski, D. J. 1957. New host records for *Meloidogyne hapla* including two plants native to California. Plant Dis. Rep. 41:770–771.

Raski, D. J., and A. M. Golden. 1965. Studies on the genus *Criconemoides* Taylor, 1936 with descriptions of eleven new species and *Bakernema variabile* n. sp. (Criconematidae:Nematoda). Nematologica 11:501–565.

Raski, D. J., and S. A. Sher. 1952. *Sphaeronema californicum*, nov. gen. nov. spec., (Criconematidae: Sphaeronematinae, nov. subfam.) an endoparasite of the roots of certain plants. Proc. Helminthol. Soc. Wash. 19:77–80.

Rau, G. J. 1963. Three new species of *Belonolaimus* (Nematoda:Tylenchidae) with additional data on *B. longicaudatus* and *B. gracilis*. Proc. Helminthol. Soc. Wash. 30:119–128.

Rebois, R. V. 1973a. Effect of soil temperature on infectivity and development of *Rotylenchulus reniformis* on resistant and susceptible soybeans, *Glycine max*. J. Nematol. 5:10–13.

Rebois, R. V. 1973b. Effect of soil water in infectivity and development of *Rotylenchulus reniformis* on soybean, *Glycine max*. J. Nematol. 5:246–249.

Reuver, I. 1959. Untersuchungen über *Paratylenchus amblycephalus* n. sp. (Nematoda, Criconematidae). Nematologica 4:3–15.

Renolds, H. W., and R. G. Hanson. 1957. Rhizoctonia disease of cotton in presence or absence of the cotton root-knot nematode in Arizona. Phytopathology 47:256–261.

Reynolds, H. W., and J. H. O'Bannon. 1966. The efficacy and residual nature of experimental chemicals for controlling *Meloidogyne incognita acrita* on Deltapine cotton in Arizona. Plant Dis. Rep. 50:512–515.

Rhoades, H. L. 1965. Parasitism and pathogenicity of *Trichodorus proximus* to St. Augustine grass. Plant Dis. Rep. 49:259–262.

Rhoades, H. L., and M. B. Linford. 1959. Control of Pythium root rot by the nematode *Aphelenchus avenae*. Plant Dis. Rep. 43:323–328.

Rhoades, H. L., and M. B. Linford. 1961a. A study of the parasitic habit of *Paratylenchus projectus* and *P. dianthus*. Proc. Helminthol. Soc. Wash. 28:185–190.

Rhoades, H. L., and M. B. Linford. 1961b. Biological studies on some members of the genus *Paratylenchus*. Proc. Helminthol. Soc. Wash. 28:51–59.

Richter, E. 1969. Zur Vertikalen verteilung von nematoden in einem Sandboden. Nematologica 15:44–54.

Riedel, R. M., and P. O. Larsen. 1974. Interrelationship of *Aphelenchoides fragariae* and *Xanthomonas begoniae* on Rieger begonia. J. Nematol. 6:215–216.

Riffle, J. W. 1963. *Meloidogyne ovalis* (Nematoda: Heteroderidae), a new species of root-knot nematode. Proc. Helminthol. Soc. Wash. 30:287–292.

Riffle, J. W. 1968. Plant-parasitic nematodes in marginal *Pinus ponderosa* stands in central New Mexico. Plant Dis. Rep. 52:52–55.

Riffle, J. W. 1969. Mycorrhizae. 8. Effect of nematodes on root-inhabiting fungi. Proc. First N. Amer. Conf. on Mycorrhizae. U.S. Dept. Agric. For. Ser. Misc-Pub. 1189:97–113.

Riffle, J. W. 1970a. *Aphelenchoides cibolensis* (Nematoda: Aphelenchoididae), a new mycophagous nematode species. Proc. Helminthol. Soc. Wash. 37:78–80.

Riffle, J. W., 1970b. Nematodes parasitic on *Pinus ponderosa*. Plant Dis. Rep. 54:752–754.

Riffle, J. W. 1972. Effect of certain nematodes on the growth of *Pinus edulis* and *Juniperus monosperma* seedlings. J. Nematol. 4:91–94.

Riffle, J. W. 1973. Effect of two mycophagous nematodes on *Armillaria mellea* root rot on *Pinus ponderosa* seedlings. Plant Dis. Rep. 57:355–357.

Riffle, J. W. 1975. Two *Aphelenchoides* species suppress formation of *Suillus granulatus* ectomycorrhizae with *Pinus ponderosa* seedlings. Plant Dis. Rep. 59:951–955.

Riffle, J. W., and J. E. Kuntz. 1967. Pathogenicity and host range of *Meloidogyne ovalis*. Phytopathology 57:104–107.

Riffle, J. W., and D. D. Lucht. 1966. Root-knot nematode on Ponderosa pine in New Mexico. Plant Dis. Rep. 50:126.

Riggs, R. D., H. Hirschmann, and M. L. Hamblen. 1969. Life cycle, host range, and reproduction of *Heterodera betulae*. J. Nematol. 1:180–183.

Riggs, R. D., D. A. Slack, and J. P. Fulton. 1956. Meadow nematode and its relation to decline of strawberry plants in Arkansas. Phytopathology 46:24. (Abstr.)

Robbins, R. T., and K. R. Barker. 1974. The effects of soil type, particle size, temperature, and moisture on reproduction of *Belonolaimus longicaudatus*. J. Nematol. 6:1–6.

Robinson, T., and A. L. Neal. 1956. The influence of hydrogen ion concentration on the emergence of golden nematode larvae. Phytopathology 46:665–667.

Rode, H. 1969. Uber verhatlen und reaktionsempfindlichkeit von larven des kartoffelnematoden gegenüber thermischen reizgefallen im überoptimalen temperaturbereich. Nematologica 15:510–524.

Rodriguez, J. G., C. F. Wade, and C. N. Wells. 1962. Nematodes as a natural food for *Macrocheles muscaedomesticae* (Acarina: Macrochelidae), a predator of the house fly egg. Ann. Entomol. Soc. Amer. 55:507–511.

Rössner, J. 1970. Ein Beitrag zur Vertikalbesiedlung des Bodens durch Wandernde Wurzelnematoden. Nematologica 16:556–562.

Rössner, J. 1971. Einfluss der Austrocknung des Bodens auf Wandernde Wurzelnematoden. Nematologica 17:127–144.

Rössner, J. 1972. Vertikalverteilung wandernder Wurzelnematoden im Boden in Abhängigkeit von Wassergehalt und Durchwurzelung. Nematologica 18:360–372.

Roggen, D. R. 1966. On the morphology of *Xiphinema index* reared on grape fanleaf virus infected grapes. Nematologica 12:287–296.

Rohde, R. A. 1960. The influence of carbon dioxide on respiration of certain plant-parasitic nematodes. Proc. Helminthol. Soc. Wash. 27:160–164.

Rohde, R. A. 1972. Expression of resistance in plants to nematodes. Ann. Rev. Phytopathol. 10:233–252.

Rohde, R. A., and W. R. Jenkins. 1957. Host range of a species of *Trichodorus* and its host–parasite relationships on tomato. Phytopathology 47:295–298.

Roman, J. 1965. Nematodes of Puerto Rico, the genus *Helicotylenchus* Steiner, 1945 (Nematoda: Hoplolaiminae). Univ. P. Rico Agric. Exp. Sta. Tech. Pap. 41. 23 p.

Ross, G. J. S., and D. L. Trudgill. 1969. The effect of population density on the sex ratio of *Heterodera rostochiensis*; a two dimensional model. Nematologica 15:601–607.

Ross, J. P. 1959. Nitrogen fertilization on the response of soybeans infected with *Heterodera glycines*. Plant Dis. Rep. 43:1284–1286.

Ross, J. P. 1969. Effect of *Heterodera glycines* on yields of nonnodulating soybeans grown at various nitrogen levels. J. Nematol. 1:40–42.

Roy, A. K. 1968. The formation of giant cells in tomato roots infected by the potato cyst nematode, *Heterodera rostochiensis* and a grey sterile fungus, G.S.F. Nematologica 14:313–314.

Ruehle, J. L. 1962. Histopathological studies of pine roots infected with lance and pine cystoid nematodes. Phytopathology 52:68–71.

Ruehle, J. L. 1966. Nematodes parasitic on forest trees. I. Reproduction of ectoparasites on pines. Nematologica 12:443–447.

Ruehle, J. L. 1967. Distribution of plant-parasitic nematodes associated with forest trees of the world. U.S. Dept. Agric. For. Ser. S.E. Forest Exp. Sta. 156 p.

Ruehle, J. L. 1968a. Pathogenicity of sting nematode on sycamore. Plant Dis. Rep. 52:523–525.

Ruehle, J. L. 1968b. Plant-parasitic nematodes associated with southern hardwood and coniferous forest trees. Plant Dis. Rep. 52:837–839.

Ruehle, J. L. 1969a. Influence of stubby-root nematode on growth of southern pine seedlings. For. Sci. 15:130–134..

Ruehle, J. L. 1969b. Nematodes parasitic on forest trees. II. Reproduction of endoparasites on pines. Nematologica 15:76–80.

Ruehle, J. L. 1971. Nematodes parasitic on forest trees: III. Reproduction on selected hardwoods. J. Nematol. 3:170–173.

Ruehle, J. L. 1972a. Nematodes of forest trees. Pp. 312–334 *in* J. M. Webster, Ed., Economic nematology. Academic, New York.

Ruehle, J. L. 1972b. Pathogenicity of *Xiphinema chambersi* on sweetgum. Phytopathology 62:333–336.

Ruehle, J. L., and J. R. Christie. 1958. Feeding and reproduction of the nematode *Hemicycliophora parvana*. Proc. Helminthol. Soc. Wash. 25:57–60.

Ruehle, J. L., and D. H. Marx. 1971. Parasitism of ectomycorrhizae of pine by lance nematode. For. Sci. 17:31–34.

Russell, R. C. 1927. A nematode discovered on wheat in Saskatchewan. Sci. Agric. 7:385–386.

Ryder, H. W., and H. W. Crittenden. 1965. Relationship of *Meloidogyne incognita acrita* and *Plasmodiophora brassicae* in cabbage roots. Phytopathology 55:506. (Abstr.)

Sabet, K. A. 1954. On the host range and systematic position of the bacteria responsible for the yellow slime disease of wheat (*Triticum vulgare* Vill.) and cocksfoot grass (*Dactylis glomerata* L.). Ann. Appl. Biol. 41:606–611.

Sandstedt, R., T. Sullivan, and M. L. Schuster. 1961. Nematode tracks in the study of movement of *Meloidogyne incognita incognita*. Nematologica 6:261–265.

Santo, G. S., and B. Lear. 1976. Influence of *Pratylenchus vulnus* and *Meloidogyne hapla* on the growth of rootstocks of rose. J. Nematol. 8:18–23.

Santos, M. S. N. de A. 1973. Mobility of males of *Meloidogyne* spp. and their responses to females. Nematologica 19:521–527.

Sanwal, K. C. 1965. Two new species of the genus *Aphelenchoides* Fischer, 1894 (Nematoda:Aphelenchoididae) from the Canadian Arctic. Can. J. Zool. 43:933–940.

Sasser, J. N. 1954. Identification and host–parasite relationships of certain root-knot nematodes (*Meloidogyne* spp.). Univ. Md. Agric. Exp. Sta. Bull. A-77. 30 p.

Sasser, J. N., G. B. Lucas, and H. R. Powers, Jr. 1955. The relationship of root-knot nematodes to black-shank resistance in tobacco. Phytopathology 45:459–461.

Sauer, M. R., and J. E. Giles. 1957. Effects of some field management systems on root knot of tomato. Nematologica 2:97–107.

Sayre, R. M. 1963. Winter survival of root-knot nematodes in southwestern Ontario. Can. J. Plant Sci. 43:361–364.

Sayre, R. M. 1964. Cold-hardiness of nematodes. I. Effects of rapid freezing on the eggs and larvae of *Meloidogyne incognita* and *M. hapla*. Nematologica 10:168–179.

Sayre, R. M. 1969. A method of culturing a predaceous tardigrade on the nematode *Panagrellus redivivus*. Trans. Amer. Microsc. Soc. 88:266–274.

Sayre, R. M. 1973. *Theratromyxa weberi*, an amoeba predatory on plant-parasitic nematodes. J. Nematol. 5:258–264.

Sayre, R. M., and S. Hwang. 1975. Freezing and storing *Ditylenchus dipsaci* in liquid nitrogen. J. Nematol. 7:199–202.

Sayre, R. M., and L. S. Keeley. 1969. Factors influencing *Catenaria anguillulae* infections in a free-living and a plant-parasitic nematode. Nematologica 15:492–502.

Sayre, R. M., and E. M. Powers. 1966. A predacious soil turbellarian that feeds on free-living and plant-parasitic nematodes. Nematologica 12:619–629.

Sayre, R. M., Z. A. Patrick, and H. J. Thorpe. 1965. Identification of a selective nematicidal component in extracts of plant residues decomposing in soil. Nematologica 11:263–268.

Schaerffenberg, B., and H. Tendl. 1951. Untersuchung über das Verhalten der Enchytraeiden gegenüber dem Zuckerrübennematoden *H. schachtii* Schm. Z. Angew. Entomol. 32:476–488.

Schindler, A. F. 1957. Parasitism and pathogenicity of *Xiphinema diversicaudatum*, an ectoparasitic nematode. Nematologica 2:25–31.

Schindler, A. F., R. N. Stewart, and P. Semeniuk. 1961. A synergistic *Fusarium*–nematode interaction in carnations. Phytopathology 51:143–146.

Schmitt, D. P. 1969. Population patterns of some stylet-bearing nematodes in a native Iowa prairie. J. Nematol. 1:304. (Abstr.)

Schmitt, D. P. 1973a. Population fluctuations of some plant parasitic nematodes in the Kalsow Prairie, Iowa. Proc. Iowa Acad. Sci. 80:69–71.

Schmitt, D. P. 1973b. Soil property influences on *Xiphinema americanum* populations as related to maturity of loess-derived soils. J. Nematol. 5:234–240.

Schmitt, D. P., and D. C. Norton. 1972. Relationships of plant parasitic nematodes to sites in native Iowa prairies. J. Nematol. 4:200–206.

Schoeneweiss, D. F., D. P. Taylor, and D. I. Edwards. 1967. Occurrence and reproduction of a stylet-bearing mycophagus nematode in pycnidia of *Cytosporina acharii*. Phytopathology 57:1392–1393.

Seinhorst, J. W. 1956. Population studies on stem eelworms (*Ditylenchus dipsaci*). Nematologica 1:159–164.

Seinhorst, J. W. 1966. The relationships between population increase and population density in plant parasitic nematodes. I. Introduction and migratory nematodes. Nematologica 12:157–169.

Seinhorst, J. W. 1967a. The relationships between population increase and population density in plant parasitic nematodes. II. Sedentary nematodes. Nematologica 13:157–171.

Seinhorst, J. W. 1967b. The relationships between population increase and population density in plant parasitic nematodes. III. Definition of the terms host, host status and resistance. IV. The influence of external conditions on the regulation of population density. Nematologica 13:429–442.

Seinhorst, J. W. 1967c. The relationships between population increase and population density in plant parasitic nematodes. V. Influence of damage to host on multiplication. Nematologica 13:481–492.

Seinhorst, J. W. 1968. Underpopulation in plant parasitic nematodes. Nematologica 14:549–553.

Seinhorst, J. W. 1970. Dynamics of populations of plant parasitic nematodes. Ann. Rev. Phytopathol. 8:131–156.

Seshadri, A. R. 1964. Investigations on the biology and life cycle of *Criconemoides xenoplax* Raski, 1952 (Nematoda:Criconematidae). Nematologica 10:540–562.

Sewell, G. W. F. 1965. The effect of altered physical condition of soil on biological control. Pp. 479–494 *in* K. F. Baker and W. C. Snyder, Eds., Ecology of soilborne plant pathogens. University of California Press, Berkeley.

Sharma, R. D. 1968. Host suitability of a number of plants for the nematode *Tylenchorhynchus dubius*. Neth. J. Plant Pathol. 74:97–100.

Shepherd, A. M. 1959. The invasion and development of some species of *Heterodera* in plants of different host status. Nematologica 4:253–267.

Shepperson, J. R., and W. C. Jordan. 1974. Observations on in vitro survival and development of *Meloidogyne*. Proc. Helminthol. Soc. Wash. 41:254.

Sher, S. A. 1959. A disease of carnation caused by the nematode *Criconemoides xenoplax*. Phytopathology 49:761–763.

Sher, S. A. 1963. Revision of the Hoplolaiminae (Nematoda). II. *Hoplolaimus* Daday, 1905 and *Aorolaimus* n. gen. Nematologica 9:267–295.

Sher, S. A. 1966. Revision of the Hoplolaiminae (Nematoda) VI. *Helicotylenchus* Steiner, 1945. Nematologica 12:1–56.

Sher, S. A. 1968. Revision of the genus *Hirschmanniella* Luc & Goodey, 1963 (Nematoda:Tylenchoidea). Nematologica 14:243–275.

Sher, S. A., and M. W. Allen. 1953. Revision of the genus *Pratylenchus*. Univ. Calif. Pub. Zool. 57:441–469.

Sher, S. A., and A. H. Bell. 1965. The effect of soil type and soil temperature on root-lesion nematode disease of roses. Plant Dis. Rep. 49:849–851.

Siddiqui, I. A., and D. P. Taylor. 1969. Feeding mechanisms of *Aphelenchoides bicaudatus* on three fungi and an alga. Nematologica 15:503–509.

Siddiqui, I. A., and D. P. Taylor. 1970a. Histopathogenesis of galls induced by *Meloidogyne naasi* in wheat roots. J. Nematol. 2:239–247.

Siddiqui, I. A., and D. P. Taylor. 1970b. The biology of *Meloidogyne naasi*. Nematologica 16:133–143.

Siddiqui, I. A., S. A. Sher, and A. M. French. 1973. Distribution of plant parasitic nematodes in California. Div. Plant Indust. Dept. Food and Agric. Sacramento. 324 p.

Sikora, R. A., D. P. Taylor, R. B. Malek, and D. I. Edwards. 1972. Interaction of *Meloidogyne naasi, Pratylenchus penetrans,* and *Tylenchorhynchus agri* on creeping bentgrass. J. Nematol. 4:162–165.

Simons, W. R. 1973. Nematode survival in relation to soil moisture. Mededel. Landbouwhogeschool Wageningen 73:1–85.

Singh, R. S., and K. Sitaramaiah. 1966. Incidence of root knot of okra and tomatoes in oil-cake amended soil. Plant Dis. Rep. 50:668–672.

Singh, R. S., B. Singh, and S. P. S. Beniwal. 1967. Observations on the effect of sawdust on incidence of root knot and on yield of okra and tomatoes in nematode-infested soil. Plant Dis. Rep. 51:861–863.

Singh, S. D. 1967. On two new species of the genus *Aphelenchoides* Fischer, 1894 (Nematoda:Aphelenchoididae) from North India. J. Helminthol. 41:63–70.

Slack, D. A., R. D. Riggs, and M. L. Hamblen. 1972. The effect of temperature and moisture on the survival of *Heterodera glycines* in the absence of a host. J. Nematol. 4:263–266.

Sledge, E. B. 1959. The extrusion of saliva from the stylet of the spiral nematode, *Helicotylenchus nannus*. Nematologica 4:356.

Sleeth, B., and H. W. Reynolds. 1955. Root-knot nematode infestation as influenced by soil texture. Soil Sci. 80:459–461.

Slootweg, A. F. G. 1956. Rootrot of bulbs caused by *Pratylenchus* and *Hoplolaimus* spp. Nematologica 1:192–201.

Smart, G. C., Jr., and H. M. Darling. 1963. Pathogenic variation and nutritional requirements of *Ditylenchus destructor*. Phytopathology 53:374–381.

Smart, G. C., Jr., and H. R. Thomas. 1969. Survival of eggs and larvae in cysts of the soybean cyst nematode, *Heterodera glycines*, ingested by swine. Proc. Helminthol. Soc. Wash. 36:139–142.

Smith, A. M. 1929. Investigations on *Heterodera schachtii* in Lancashire and Cheshire. Part II. The relationships between degree of infestation and hygroscopic moisture, loss on ignition and pH value of the soil. Ann. Appl. Biol. 16:340–346.

Smolik, J. D. 1974. Nematode studies at the Cottonwood site. U.S. IBP Grassland Biome Tech. Rep. No. 251. Colorado State University, Fort Collins. 80 p.

Smolik, J. D., and L. E. Rogers. 1976. Effects of cattle grazing and wildfire on soil-dwelling nematodes of the shrub-steppe ecosystem. J. Range Management 29:304–306.

Sneath, P. H. A., and R. R. Sokal. 1973. Numerical taxonomy. Freeman, San Francisco.

Sol, H. H., and J. W. Seinhorst. 1961. The transmission of rattle virus by *Trichodorus pachydermus*. Tijdsch. Plantenziek. 67:307–309.

Soprunov, F. F. 1958. Predacious Hyphomycetes and their application in the control of pathogenic nematodes. Trans. from Russian. Israel Prog. for Scientific Translations 1966. U.S. Dept. Commerce. 292 p.

Sorensen, T. 1948. A method of establishing groups of equal amplitude in plant sociology based on similarity of species content and its application to analyses of the vegetation on Danish Commons. K. Danske Biol. Skr. Copenhagen 5(4):1–34.

Southards, C. J. 1971. Effect of fall tillage and selected hosts on the population density of *Meloidogyne incognita* and *Pratylenchus zeae*. Plant Dis. Rep. 55:41–44.

Southey, J. F. 1956. National survey work for cereal root eelworm (*Heterodera major* (O. Schmidt) Franklin)). Nematologica 1:64–71.

Southey, J. F. 1974. Methods for detection of potato cyst nematodes. EPPO Bull. 4:463–473.

Springer, J. K. 1964. Nematodes associated with plants in cultivated woody plant nurseries and uncultivated woodland areas in New Jersey. New Jersey Dept. Agr. Circ. 429. 40 p.

Starr, J. L., and W. F. Mai. 1976a. Effect of soil microflora on the interaction of three plant-parasitic nematodes with celery. Phytopathology 66:1224–1228.

Starr, J. L., and W. F. Mai. 1976b. Predicting onset of egg production by *Meloidogyne hapla* on lettuce from field soil temperatures. J. Nematol. 8:87–88.

Steele, A. E. 1975. Population dynamics of *Heterodera schachtii* on tomato and sugarbeet. J. Nematol. 7:105–111.

Steiner, G. 1916. Freilebende nematoden von Nowaja-Semlja. Zool. Anz. 47:50–74.

Steiner, G. 1923. Intersexes in nematodes. J. Heredity 14:147–158.

Steiner, G. 1931. *Neotylenchus abulbosus* n. g., n. sp. (Tylenchidae, Nematoda) the causal agent of a new nematosis of various crop plants. J. Wash. Acad. Sci. 21:536–538.

Steiner, G. 1933. *Rhabditis lambdiensis*, a nematode possibly acting as a disease agent in mushroom beds. J. Agric. Res. 46:427–435.

Steiner, G. 1936. Opuscula miscellanea nematologica, IV. Proc. Helminthol. Soc. Wash. 3:74–80.

Steiner, G. 1937. Opuscula miscellanea nematologica, V. Proc. Helminthol. Soc. Wash. 4:33–38.

Steiner, G., and F. M. Albin. 1946. Resuscitation of the nematode *Tylenchus polyhypnus*, n. sp., after almost 39 years' dormancy. J. Wash. Acad. Sci. 36:97–99.

Steiner, G., and E. M. Buhrer. 1933. Recent observations on diseases caused by nematodes. Plant Dis. Rep. 17:172–173.

Steiner, G., and E. M. Buhrer. 1934. *Aphelenchoides xylophilus*, n. sp., a nematode associated with blue-stain and other fungi in timber. J. Agric. Res. 48:949–951.

Steiner, G., and H. Heinly. 1922. The possibility of control of *Heterodera radicicola* and other plant-injurious nemas by means of predatory nemas, especially by *Mononchus papillatus* Bastian. J. Wash. Acad. Sci. 12:367–386.

Steiner, G., and C. E. Scott. 1934. A nematosis of *Amsinckia* caused by a new variety of *Anguillulina dipsaci*. J. Agric. Res. 49:1087–1092.

Stephenson, W. 1945. The effect of acids on a soil nematode. Parasitology 36:158–164.

Stessel, G. J., and A. M. Golden. 1961. Occurrence of *Ditylenchus radicicola* (Nematoda: Tylenchidae) in the United States and on a new host. Plant Dis. Rep. 45:26–28.

Stewart, R. N., and A. F. Schindler. 1956. The effect of some ectoparasitic and endoparasitic nematodes on the expression of bacterial wilt of carnations. Phytopathology 46:219–222.

Streu, H. T., W. R. Jenkins, and M. T. Hutchinson. 1961. Nematodes associated with carnations *Dianthus caryophyllus* L. with special reference to the parasitism and biology of *Criconemoides curvatum* Raski. New Jersey Agric. Exp. Sta. Bull. 800. 32 p.

Suit, R. F., E. P. Du Charme, T. L. Brooks, and H. W. Ford. 1953. Factors in the control of the burrowing nematode on citrus. Proc. Fla. State Hort. Soc. 66:46–49.

Sutherland, J. R. 1967a. Host range and reproduction of the nematodes *Paratylenchus projectus, Pratylenchus penetrans,* and *Tylenchus emarginatus* on some forest nursery seedlings. Plant Dis. Rep. 51:91–93.

Sutherland, J. R. 1967b. Parasitism of *Tylenchus emarginatus* on conifer seedling roots and some observations on the biology of the nematode. Nematologica 13:191–196.

Sutherland, J. R. 1969. Feeding of *Xiphinema bakeri*. Phytopathology 59:1963–1965.

Sutherland, J. R., and R. E. Adams. 1964. The parasitism of red pine and other forest nursery crops by *Tylenchorhynchus claytoni* Steiner. Nematologica 10:637–643.

Sutherland, J. R., and R. E. Adams. 1966. Population fluctuations of nematodes associated with red pine seedlings following chemical treatment of the soil. Nematologica 12:122–128.

Sutherland, J. R., and J. A. Fortin. 1968. Effect of the nematode *Aphelenchus avenae* on some ectotrophic mycorrhizal fungi and on a red pine mycorrhizal relationship. Phytopathology 58:519–523.

Sutherland, J. R., and D. A. Ross. 1971. Temperature effects on survival of *Xiphinema bakeri* in fallow soil. J. Nematol. 3:276–279.

Sutherland, J. R., and L. J. Sluggett. 1974. Time, temperature, and soil moisture effects on *Xiphinema bakeri* nematode survival in fallow soil. Phytopathology 64:507–513.

Taha, A. H. Y., and D. J. Raski. 1969. Interrelationships between root-nodule bacteria, plant-parasitic nematodes and their leguminous host. J. Nematol. 1:201–211.

Tarjan, A. C. 1960. Longevity of the burrowing nematode, *Radopholus similis*, in host-free soil. Phytopathology 50:656–657. (Abstr.)

Tarjan, A. C. 1961. Attempts at controlling citrus burrowing nematodes using nematode-trapping fungi. Soil and Crop Sci. Soc. Fla. 21:17–36.

Taubenhaus, J. J., and W. N. Ezekiel. 1933. Checklist of diseases of plants in Texas. Trans. Texas Acad. Sci. 16:5–118.

Taylor, A. L. 1935. A review of the fossil nematodes. Proc. Helminthol. Soc. Wash. 2:47–49.

Taylor, A. L., and E. M. Buhrer. 1958. A preliminary report on distribution of root-knot nematode species in the United States. Phytopathology 48:464. (Abstr.)

Taylor, C. E. 1971. Nematodes as vectors of plant viruses. Pp. 185–211 *in* B. M. Zuckerman, W. F. Mai, and R. A. Rohde, Eds., Plant parasitic nematodes. Vol. II. Academic, New York.

Taylor, C. E., and A. F. Murant. 1966. Nematicidal activity of aqueous extracts from raspberry canes and roots Nematologica 12:488–494.

Taylor, C. E., and D. J. Raski. 1964. On the transmission of grape fanleaf by *Xiphinema index*. Nematologica 10:489–495.

Taylor, D. P. 1961. Biology and host–parasite relationships of the spiral nematode, *Helicotylenchus microlobus*. Proc. Helminthol. Soc. Wash. 28:60–66.

Taylor, D. P., and W. R. Jenkins. 1957. Variation within the nematode genus *Pratylenchus*, with the descriptions of *P. hexincisus*, n. sp. and *P. subpenetrans*, n. sp. Nematologica 2:159–174.

Thames, W. H., Jr., and W. N. Stoner, 1953. A preliminary trial of lowland culture rice in rotation with vegetable crops as a means of reducing root-knot nematode infestations in the Everglades. Plant Dis. Rep. 37:187–192.

Thomas, H. A. 1959. On *Criconemoides xenoplax* Raski, with special reference to its biology under laboratory conditions. Proc. Helminthol. Soc. Wash. 26:55–59.

Thomas, S. H. 1978. Population densities of nematodes under seven tillage regimes. J. Nematol. 10:24–27.

Thomason, I. J. 1959. Influence of soil texture on development of the stubby-root nematode. Phytopathology 49:552. (Abstr.)

Thomason, I. J., and D. Fife. 1962. The effect of temperature on development and survival of *Heterodera schachtii* Schm. Nematologica 7:139–145.

Thomason, I. J., and B. Lear. 1961. Rate of reproduction of *Meloidogyne* spp. as influenced by soil temperature. Phytopathology 51:520–524.

Thomason, I. J., and F. C. O'Melia. 1962. Pathogenicity of *Pratylenchus scribneri* of crop plants. Phytopathology 52:755. (Abstr.)

Thomason, I. J., D. C. Erwin, and M. J. Garber. 1959. The relationship of the root-knot nematode, *Meloidogyne javanica*, to Fusarium wilt of cowpea. Phytopathology 49:602–606. ·

Thorne, G. 1926a. Control of sugar-beet nematode by crop rotation. U.S. Dept. Agric. Farmers Bull. 1514. 20 p.

Thorne, G. 1926b. *Tylenchus balsamophilus*, a new plant parasitic nematode. J. Parasitol. 12:141–145.

Thorne, G. 1927. The life history, habits, and economic importance of some mononchs. J. Agric. Res. 34:265–286.

Thorne, G. 1928. *Heterodera punctata* n. sp., a nematode parasitic on wheat roots from Saskatchewan. Sci. Agric. 8:707–711.

Thorne, G. 1929. Nematodes from the summit of Long's Peak, Colorado. Trans. Amer. Microsc. Soc. 48:181–195.

Thorne, G. 1930. Predacious nemas of the genus *Nygolaimus* and a new genus, *Sectonema*. J. Agric. Res. 41:445–466.

Thorne, G. 1934. Some plant-parasitic nemas, with descriptions of three new species. J. Agric. Res. 49:755–763.

Thorne, G. 1939. A monograph of the nematodes of the superfamily Dorylaimoidea. Capita Zool. 8(5):1–261.

Thorne, G. 1940. *Duboscqia penetrans*, n. sp. (Sporozoa, Microsporidia, Nosematidae), a parasite of the nematode *Pratylenchus pratensis* (De Man) Filipjev. Proc. Helminthol. Soc. Wash. 7:51–53.

Thorne, G. 1941. Some nematodes of the family Tylenchidae which do not possess a valvular median esophageal bulb. Great Basin Natural. 2:37–85.

Thorne, G. 1943. *Cacopaurus pestis*, nov., gen., nov. spec. (Nematoda:Criconematinae), a destructive parasite of the walnut, *Juglans regia*. Linn. Proc. Helminthol. Soc. Wash. 10:78–83.

Thorne, G. 1949. On the classification of the Tylenchida, a new order (Nematoda, Phasmidia). Proc. Helminthol. Soc. Wash. 16:37–73.

Thorne, G. 1955. Fifteen new species of the genus *Hemicycliophora* with an amended description of *H. typica* de Man (Tylenchida Criconematidae). Proc. Heminthol. Soc. Wash. 22:1–16.

Thorne, G. 1961. Principles of nematology. McGraw-Hill, New York.

Thorne, G. 1974. Nematodes of the northern Great Plains. Part II. Dorylaimoidea in part (Nemata:Adenophorea). S. Dak. Agric. Exp. Sta. Tech. Bull. 41. 120 p.

Thorne, G., and R. B. Malek. 1968. Nematodes of the Northern Great Plains. S. Dak. Agric. Exp. Sta. Tech. Bull. 31. 111 p.

Thorne, G., and M. L. Schuster. 1956. *Nacobbus batatiformis*, n. sp. (Nematoda:Tylenchidae) producing galls on the roots of sugar beets and other plants. Proc. Helminthol. Soc. Wash. 23:128–134.

Thorne, G., and H. H. Swanger. 1936. A monograph of the nematode genera *Dorylaimus* Dujardin, *Aporcelaimus* n. g., *Dorylaimoides* n. g. and *Pungentus* n. g. Capita Zool. 6(4):1–223.

Tikyani, M. G., and S. Khera. 1969a. In vitro feeding of *Telotylenchus loofi* on great millet (*Sorghum vulgare*) roots Nematologica 15:291–293.

Tikyani, M. G., and S. Khera. 1969b. Studies on the culturing of certain Aphelenchs. Indian J. Helminthol. 21:6–12.

Tjepkema, J. P., J. A. Knierim, and N. Knobloch. 1967. The plant-parasitic nematodes associated with cultivated blueberries in Michigan. Mich. Agric. Exp. Sta. Quart. Bull. 50:37–45.

Townshend, J. L. 1964. Fungus hosts of *Aphelenchus avenae* Bastian, 1865 and *Bursaphelenchus fungivorus* Franklin & Hooper, 1962 and their attractiveness to these nematode species. Can. J. Microbiol. 10:727–737.

Townshend, J. L. 1972. Influence of edaphic factors on penetration of corn roots by *Pratylenchus penetrans* and *P. minyus* in three Ontario soils. Nematologica 18:201–212.

Townshend, J. L. 1973. Survival of *Pratylenchus penetrans* and *P. minyus* in two Ontario soils. Nematologica 19:35–42.

Townshend, J. L., and R. E. Blackith. 1975. Fungal diet and the morphometric relationships in *Aphelenchus avenae*. Nematologica 21:19–25.

Townshend, J. L., and L. R. Webber. 1971. Movement of *Pratylenchus penetrans* and the moisture characteristics of three Ontario soils. Nematologica 17:47–57.

Transeau, E. N. 1935. The prairie peninsula. Ecology 16:423–437.

Triantaphyllou, A. C. 1960. Sex determination in *Meloidogyne incognita* Chitwood, 1949 and intersexuality in *M. javanica* (Treub, 1885) Chitwood, 1949. Ann. Inst. Phytopathol. Benaki (N.S.) 3:12–31.

Trudgill, D. L. 1967. The effect of environment on sex determination in *Heterodera rostochiensis*. Nematologica 13:263–272.

Trudgill, D. L. 1970. Survival of different stages of *Heterodera rostochiensis* at high temperatures. Nematologica 16:94–98.

Trudgill, D. L. 1976. Observations on the feeding of *Xiphinema diversicaudatum*. Nematologica 22:417–423.

Tseng, S-T., K. R. Allred, and G. D. Griffin. 1968. A soil population study of *Ditylenchus dipsaci* (Kühn) Filipjev in an alfalfa field. Proc. Helminthol. Soc. Wash. 35:57–62.

Tu, C. C., and Y. H. Cheng. 1971. Interaction of *Meloidogyne javanica* and *Macrophomina phaseoli* in Kenaf root rot. J. Nematol. 3:39–42.

Turner, D. R., and R. A. Chapman. 1972. Infection of seedlings of alfalfa and red clover by concomitant populations of *Meloidogyne incognita* and *Pratylenchus penetrans*. J. Nematol. 4:280–286.

Tyler, J. 1933a. Development of the root-knot nematode as affected by temperature. Hilgardia 7:391–415.

Tyler, J. 1933b. Reproduction without males in aseptic root cultures of the root-knot nematode. Hilgardia 7:373–388.

Tyler, J. 1938. Egg output of the root-knot nematode. Proc. Helminthol. Soc. Wash. 5:49–54.

Upadhyay, K. D., and G. Swarup. 1972. Culturing, host range and factors affecting multiplication of *Tylenchorhynchus vulgaris* on maize. Indian J. Nematol. 2:139–145.

van der Laan, P. A. 1954. Nader onderzoek over het aaltjesvangende amoeboide organisme *Theratromyxa weberi* Zwillenberg. Tijdschr. Plziekt. 60:139–145.

Van Gundy, S. D. 1958. The life history of the citrus nematode *Tylenchulus seminpenetrans* Cobb. Nematologica 3:283–294.

Van Gundy, S. D. 1965. Factors in survival of nematodes. Ann. Rev. Phytopathol. 3:43–68.

Van Gundy, S. D., and J. P. Martin. 1961. Influence of *Tylenchulus semipenetrans* on the growth and chemical composition of sweet orange seedlings in soils of various exchangeable cation ratios. Phytopathology 51:146–151.

Van Gundy, S. D., and R. L. Rackham. 1961. Studies on the biology and pathogenicity of *Hemicycliophora arenaria*. Phytopathology 51:393–397.

Van Gundy, S. D., and L. H. Stolzy. 1963. The relationship of oxygen diffusion rates to the survival, movement, and reproduction of *Hemicycliophora arenaria*. Nematologica 9:605–612.

Van Gundy, S. D., and P. H. Tsao. 1963. Growth reduction of citrus seedlings by *Fusarium solani* as influenced by the citrus nematode and other soil factors. Phytopathology 53:488–489.

Van Gundy, S. D., L. H. Stolzy, T. E. Szsuszkiewicz, and R. L. Rackham. 1962. Influence of oxygen supply on survival of plant parasitic nematodes in soil. Phytopathology. 52:628–632.

Van Gundy, S. D., J. P. Martin, and P. H. Tsao. 1964. Some soil factors influencing reproduction of the citrus nematode and growth reduction of sweet orange seedlings. Phytopathology 54:294–299.

Van Gundy, S. D., A. F. Bird, and H. R. Wallace. 1967. Aging and starvation in larvae of *Meloidogyne javanica* and *Tylenchulus semipenetrans*. Phytopathology 57:559–571.

Van Gundy, S. D., F. D. McElroy, A. F. Cooper, and L. H. Stolzy. 1968. Influence of soil temperature, irrigation and aeration on *Hemicycliophora arenaria*. Soil Sci. 106:270–274.

Vanterpool, T. C. 1948. *Ditylenchus radicicola* (Greeff) Filipjev, a root-gall nematode new to Canada, found on wheat and other Gramineae. Sci. Agric. 28:200–205.

Vargas, J. M., Jr., and C. W. Laughlin. 1972. The role of *Tylenchorhynchus dubius* in the development of Fusarium blight of Merion Kentucky bluegrass. Phytopathology 62:1311–1314.

Vasudeva, R. S., and M. K. Hingorani. 1952. Bacterial disease of wheat caused by *Corynebacterium tritici* (Hutchinson) Bergey et al. Phytopathology 42:291–293.

Viglierchio, D. R. 1961a. Attraction of parasitic nematodes by plant root emanations. Phytopathology 51:136–142.

Viglierchio, D. R. 1961b. Effects of storage environment on "in vitro" hatching of larvae from cysts of *Heterodera schachtii* Schmidt, 1871. Phytopathology 51:623–625.

Viglierchio, D. R., N. A. Croll, and J. H. Gortz. 1969. The physiological response of nematodes to osmotic stress and an osmotic treatment for separating nematodes. Nematologica 15:15–21.

Volz, P. 1951. Untersuchungen über die Mikrofauna des Waldbodens. Zool. Jb. (Syst.) 79:514–566.

Walker, J. T. 1969. *Pratylenchus penetrans* (Cobb) populations as influenced by microorganism and soil amendments. J. Nematol. 1:260–264.

Walker, J. T. 1971. Populations of *Pratylenchus penetrans* relative to decomposing nitrogenous soil amendments. J. Nematol. 3:43–49.

Walker, J. T., C. H. Specht, and S. Mavrodineau. 1967. Reduction of lesion nematodes in soybean meal and oil-amended soils. Plant Dis. Rep. 51:1021–1024.

Wallace, H. R. 1958a. Movement of eelworms. I. The influence of pore size and moisture content of the soil on the migration of larvae of the beet eelworm, *Heterodera schachtii* Schmidt. Ann. Appl. Biol. 46:74–85.

Wallace, H. R. 1958b. Movement of eelworm. II. A comparative study of the movement in soil of *Heterodera schachtii* Schmidt and of *Ditylenchus dipsaci* (Kuhn) Filipjev. Ann. Appl. Biol. 46:86–94.

Wallace, H. R. 1958c. Movement of eelworms. III. The relationship between eelworm length, activity and mobility. Ann. Appl. Biol. 46:662–668.

Wallace, H. R. 1959a. Further observations on some factors influencing the emergence of larvae from cysts of the beet eelworm, *Heterodera schachtii* Schmidt. Nematologica 4:245–252.

Wallace, H. R. 1959b. Movement of eelworms. V. Observations on *Aphelenchoides ritzema-bosi* (Schwartz, 1912) Steiner, 1932 on florists' chrysanthemums. Ann. Appl. Biol. 47:350–360.

Wallace, H. R. 1961. The orientation of *Ditylenchus dipsaci* to physical stimuli. Nematologica 6:222–236.

Wallace, H. R. 1962. Observations on the behaviour of *Ditylenchus dipsaci* in soil. Nematologica 7:91–101.

Wallace, H. R. 1963. The biology of plant parasitic nematodes. Arnold, London.

Wallace, H. R. 1968a. The influence of aeration on survival and hatch of *Meloidogyne javanica*. Nematologica 14:223–230.

Wallace, H. R. 1968b. The influence of soil moisture on survival and hatch of *Meloidogyne javanica*. Nematologica 14:231–242.

Wallace, H. R. 1969. The influence of nematode numbers and of soil particle size, nutrients and temperature on the reproduction of *Meloidogyne javanica*. Nematologica 15:55–64.

Wallace, H. R. 1971a. Abiotic influences in the soil environment. Pp. 257–280 *in* B. M. Zuckerman, W. F. Mai, and R. A. Rohde, Eds., Plant parasitic nematodes. Academic, New York.

Wallace, H. R. 1971b. The influence of temperature on embryonic development and hatch in *Meloidogyne javanica*. Nematologica 17:179–186.

Wallace, H. R. 1973. Nematode ecology and plant disease. Arnold, London.

Wallace, H. R., and D. N. Greet. 1964. Observations on the taxonomy and biology of *Tylenchorhynchus macrurus* (Goodey, 1932) Filipjev, 1936 and *Tylenchorhynchus icarus* sp. nov. Parasitology 54:129–144.

Wasilewska, L. 1971. Nematodes of the dunes in the Kampinos forest. II. Community structure based on numbers of individuals, state of biomass and respiratory metabolism. Ekologia Polska 19:651–688.

Wasilewska, L. 1974. Number, biomass and metabolic activity of nematodes of two cultivated fields in Turew. Zeszyty Prob. Post. nauk Rolnic. 154:419–442.

Wasilewska, L. 1975. Quantitative distribution and respiratory metabolisms with suggestion on production of nematodes on mountain pastures. Proc. 5th International Colloquium Soil·Zool. Prague 1973:133–140.

Watson, J. R. 1921. Control of root-knot, II. Fla. Agric. Expt. Sta. Bull. 159:30–44.

Watson, T. R., and B. F. Lownsbery. 1970. Factors influencing the hatching of *Meloidogyne naasi*, and a comparison with *M. hapla*. Phytopathology 60:457–460.

Weber, A. P., L. O. Zwillenberg, and P. A. van der Laan. 1952. A predacious amoeboid organism destroying larvae of the potato root eelworm and other nematodes. Nature 169:384–385.

Webley, D. 1974. An early record of *Heterodera humili* Filipjev in the United Kingdom. Nematologica 20:262.

Webster, J. M. 1964. Population increase of *Ditylenchus dipsaci* (Kuhn) in the narcissus and the spread of the nematode through the soil. Ann. Appl. Biol. 53:485–492.

Webster, J. M., and D. N. Greet. 1967. The effect of a host crop and cultivations on the rate that *Ditylenchus dipsaci* reinfested a partially sterilized area of land. Nematologica 13:295–300.

Weischer, B. 1969. Vermehrung und Shadwirkung von *Aphelenchoides ritzemabosi* und *Ditylenchus dipsaci* in virusfreiem und in TMV infiziertem Tabak. Nematologica 15:334–336.

White, J. H. 1953. Wind-borne dispersal of potato-root eelworm. Nature 172:686–687.

Whitney, E. D. 1974. Synergistic effect of *Pythium ultimum* and the additive effect of *P. aphanidermatum* with *Heterodera schachtii* on sugarbeet. Phytopathology 64:380–383.

Wieser, W. 1955. The attractiveness of plants to larvae of root-knot nematodes. I. The effect of tomato seedlings and excised roots on *Meloidogyne hapla* Chitwood. Proc. Helminthol. Soc. Wash. 22:106–112.

Wieser, W. 1956. The attractiveness of plants to larvae of root-knot nematodes. II. The effect of excised bean, eggplant, and soybean roots on *Meloidogyne hapla* Chitwood. Proc. Helminthol. Soc. Wash. 23:59–64.

Williams, J. R. 1960. Studies on the nematode soil fauna of sugarcane fields in Mauritius. 5. Notes upon a parasite of root-knot nematodes. Nematologica 5:37–42.

Williams, J. R. 1967. Observations on parasitic protozoa in plant-parasitic and free-living nematodes. Nematologica 13:336–342.

Williams, W. T., and J. M. Lambert. 1959. Multivariate methods in plant ecology. I. Association-analysis in plant communities. J. Ecol. 47:83–101.

Willis, C. B. 1972. Effects of soil pH on reproduction of *Pratylenchus penetrans* and forage yield of alfalfa. J. Nematol. 4:291–295.

Wilson, E. O., and W. H. Bossert. 1971. A primer of population biology. Sinauer Assoc., Stamford, Conn.

Winogradsky, S. 1928. The direct methods in soil microbiology and its application to the study of nitrogen fixation. Soil. Sci. 25:37–43.

Winslow, R. D. 1960. Some aspects of the ecology of free-living and plant-parasitic nematodes. Pp. 341–415 in J. N. Sasser and W. R. Jenkins, Eds., Nematology. University of North Carolina Press, Chapel Hill.

Winslow, R. D. 1964. Soil nematode population studies. I. The migratory root Tylenchida and other nematodes of the Rothamsted and Woburn six-course rotations. Pedobiologica 4:65–76.

Winslow, R. D., and T. D. Williams. 1957. Ameboid organisms attacking larvae of the potato root eelworm (*Heterodera rostochiensis* Woll.) in England and the beet eelworm (*H. schachtii* Schm.) in Canada. Tijdsch. Plantez. 63:242–243.

Wolfe, J. N. 1951. The glacial border—climatic, soil and biotic features. The possible role of microclimate. Ohio J. Sci. 51:134–138.

Wong, K., and J. M. Ferris. 1968. Factors influencing the population fluctuation of *Pratylenchus penetrans* in soil. III. Host plant species. Phytopathology 58:662–665.

Wong, T. K., and W. F. Mai. 1973. *Meloidogyne hapla* in organic soil: Effects of environment on hatch, movement, and root invasion. J. Nematol. 5:130–138.

Wood, F. H. 1974. Biology of *Seinura demani* (Nematoda: Aphelenchoididae). Nematologica 20:347–353.

Wright, H. E., Jr., and D. G. Frey, Eds. 1965. The Quaternary of the United States. Princeton University Press, Princeton, N.J.

Wu, L-Y. 1965. Five new species of *Criconemoides* Taylor, 1936 (Criconematidae: Nematoda) from Canada. Can. J. Zool. 43:203–214.

Wu, L-Y. 1967. *Anguina calamagrostis*, a new species from grass, with an emendation of the generic characters for the genera *Anguina* Scopoli, 1777 and *Ditylenchus* Filipjev, 1936 (Tylenchidae: Nematoda). Can. J. Zool. 45:1003–1010.

Wu, L-Y. 1969a. Five new species of *Tylenchus* Bastian, 1865 (Nematoda: Tylenchidae) from the Canadian high Arctic. Can. J. Zool. 47:1005–1010.

Wu, L-Y. 1969b. Dactylotylenchinae, a new subfamily (Tylenchidae: Nematoda). Can. J. Zool. 47:909–911.

Wuest, P. J., and J. R. Bloom. 1965. Effect of temperature and age of egg population on the in vitro hatching of *Meloidogyne hapla* eggs. Phytopathology 55:885–888.

Wyss, U. 1970a. Untersuchungen zur populationsdynamik von *Longidorus elongatus*. Nematologica 16:74–84.

Wyss, U. 1970b. Zur toleranz wandernder wurzelnematoden gegenüber zunehmender austrocknung des Bodens und hohen osmotischen drucken. Nematologica 16:63–73.

Yassin, A. M. 1969. Glass-house and laboratory studies on the biology of the needle nematode, *Longidorus elongatus*. Nematologica 15:169–178.

Yeates, G. W. 1968. An analysis of annual variation of the nematode fauna in dune sand, at Himatangi Beach, New Zealand. Pedobiologia 8:173–207.

Yeates, G. W. 1970. The diversity of soil nematode faunas. Pedobiologia. 10:104–107.

Yeates, G. W. 1973. Morphometrics and growth in eight New Zealand soil nematode populations. New Zealand J. Sci. 16:711–725.

Yeates, G. W. 1976. Effect of fertiliser treatment and stocking rate on pasture nematode populations in a yellow-grey earth. N. Zealand J. Agric. Res. 19:405–408.

Yuen, P-H. 1966. The nematode fauna of the regenerated woodland and grassland of Broadbalk Wilderness. Nematologica 12:195–214.

Yuksel, H. S. 1960. Observations on the life cycle of *Ditylenchus dipsaci* on onion seedlings. Nematologica 5:289–296.

Zak, B. 1967. A nematode (*Meloidodera* sp.) on Douglas-fir mycorrhizae. Plant Dis. Rep. 51:264.

Zak, B., and H. J. Jensen. 1975. Two new hosts of *Nacobbodera chitwoodi*. Plant Dis. Rep. 59:232.

Ziegler, H. E. 1895. Untersuchungen über die ersten Entwicklungsvorgänge der Nematoden. Zeitsch. Wiss. Zool. 60:351–410.

Zuckerman, B. M. 1960. Parasitism of cranberry roots by *Tetylenchus joctus* Thorne. Nematologica 5:253–254.

Zuckerman, B. M. 1961. Parasitism and pathogenesis of the cultivated cranberry by some nematodes. Nematologica 6:135–143.

Zuckerman, B. M. 1964. Studies of two nematode species associated with roots of the cultivated highbush blueberry. Plant Dis. Rep. 48:170–172.

Zuckerman, B. M., and J. W. Coughlin. 1960. Nematodes associated with some crop plants in Massachusetts. Univ. Mass. Agric. Exp. Sta. Bull. 521. 18 p.

Zuckerman, B. M., S. Khera, and A. R. Pierce. 1964. Population dynamics of nematodes in cranberry soils. Phytopathology 54:654–659.

Zullini, A. 1970. I nematodi muscicoli della val Zebru (Parco Nazionale Dello Stelvia). Instituto Lombardo Acad. di Sci. e Lett., Milan. Rendiconti B. 104:88–137.

Zullini, A. 1971. Studio sulle variazioni del popolamento nematologico in un muschio. Instituto Lombardo Acad. di Sci. e Lett., Milan. Rendiconti B. 105:89–106.

INDEX